STEP BY STEP
MIXING

HOW TO CREATE GREAT MIXES USING
ONLY 5 PLUG-INS

Björgvin Benediktsson

Step By Step Mixing
How to Create Great Mixes Using Only 5 Plug-ins

Paperback ISBN: 978-1-7336888-0-2

Published by
Björgvin Benediktsson & Audio Issues
Tucson, Arizona

www.Audio-Issues.com

More about this book and additional resources can be found at
www.StepByStepMixing.com/Resources.

Table of Contents

Free Gift: The Step By Step Mixing Resources

As a way of saying thank you for your purchase I want to give you exclusive access to the *Step by Step Mixing* resources: www.StepByStepMixing.com/Resources

Here's what you get inside the *Step By Step Mixing* resource page:

- Step By Step Mixing Quickstart video
- Mix Translation Cheatsheet
- Links to over 100 multi-tracks to practice your mixes
- Step By Step Mastering Guide
- In-Depth Frequency Chart for the Entire EQ Spectrum

Teaching others how to make a bigger impact with their music and audio production is a passion of mine. I created Audio Issues in order to pay my knowledge forward and help you make better sounding music.

You'll also get access to more information about all the resources, plug-ins and equipment I talk about throughout the book, as well as any important updates you need to know about.

Access the free resource page here:
www.StepByStepMixing.com/Resources

Introduction

Whenever I talk to home studio producers, they always tell me about the same issues when they're mixing.

It goes something like this:

> *"I have a hard time understanding the frequency spectrum when I'm trying to make all the instruments fit together in my mix. I can't figure out how to make my guitars sound the way I hear them on the records I like and they always seem to mask the vocals.*
>
> *The kick drum and the bass guitar always seem to be fighting in the low-end, and even when I use cool tricks I learned on the internet the lows are still muddy. Every time I open up a compressor I get nervous about all the different knobs and buttons. I always end up squashing my drums too hard and my mix just feels lifeless and a long cry from the dynamic mixes I hear my favorite mixing engineers release.*
>
> *Even when I manage to get my instruments sounding like they belong together I screw it all up as soon as I start adding effects like reverb and delay to the mix. And last but not least, I usually just leave the saturation alone, even though I know it's supposed to add*

something cool to my mix. I'm just too nervous and afraid of distorting my mix."

Sound familiar?

Of course it does. That's why you're reading this book.

But let me tell you a secret. Even though you can mix a song a million different ways (that's an exaggeration, but you get the picture), you can get to a quality mix quite easily by just using 20% of the processors and plug-ins available to you.

What 20% you ask? Well, the five most important processors that every skilled mixing engineer uses:

- EQ
- Compression
- Reverb
- Delay
- Saturation

These are the 20% of the plug-ins you use to get 80% (and beyond) of the results of a great mix. I would even go so far to say that the last 20% to get to a 100% awesome mix is in the recordings themselves, not in some fancy premium plug-ins (although they are very fun to play around with).

The fact is, if you master these five processors above you'll be ready to make a killer mix in your home studio, whether you're working on demos for your band or mixing records for your friends and clients.

It's my hope that after you've read this book, you'll share your successes with me, like Greg did a while ago:

"Hey brother! I have your [Step By Step Mixing book]. I purchased it about 1 1/2 years ago and today I'm

getting booked to mix tons of people's tracks now! So thankful to find your stuff man!!! I'm a producer but was always having trouble getting my tracks to sound right so I would always hire someone to mix it for me and it turned out ok but never the way I wanted it to since I had a vision and I could never get it out of my head since I didn't know the tools! Anyways man! You've made a world of difference in my income bro! So thankful!"

So in the following pages we'll dive deep into these five plug-ins and I'll help you learn what all the fuss is about.

Each chapter will have two parts:

1. **Explanations on the theory behind what the plug-ins do and how to use them:** You'll get a thorough walkthrough of the various regions of the frequency spectrum. You'll understand exactly how to use your compressors. You'll learn all about the various reverb and delay settings (some reverbs are just too complex!) and you'll get very familiar with using saturation (without overloading and distorting your mixes!).

2. **Common Problems and Their Practical Solutions:** After you understand how each processor works, we'll talk about some real-world scenarios. I'll give you some practical and easy-to-use tips to make your mixes jump out of the speakers.

How's that for hyperbole? Ready to get started?

Let's go!

A Note About Mixing

Mixing is a very subjective concept. There are a lot of varia-tions of a "good" mix. A mix you pay $200 for and a mix you pay $400 for (or even $4000) isn't always double the quality. It has to do with the experience of the mixing engineer, how in-demand their services are, and their willingness to work within whatever budget you have.

In addition, mixing is hard to teach because each song is different and poses different challenges. In the following pages I will be giving you my "80/20 rules" for mixing. It's all about what you should focus on the most in order to get the biggest wins. You'll be learning about the most important mixing subjects so you can mix your own music with confidence.

Remember, mixing skills improve with practice, not from reading books (ironically enough...) so I encourage you to mix as much music as you can in order to improve your skills. If you're looking for songs to mix, I have plenty of resources for you, with a link to hundreds of multi-tracks in the *Step By Step Mixing* resource section: www.StepByStepMixing.com/Resources.

I'm certainly not the best mixing engineer in the world. Sometimes I don't even think I'm that good, but that's more to do with the everlasting presence of the "imposter syndrome" and my neurotic insecurity than actual lack of skill. However, I do know enough to teach you the things to focus on and the mistakes to avoid. I'm constantly learning with every new production and you should as well.

So read this book with an emphasis on keeping the big ideas in mind and trying everything out on as many multi-tracks as you can get your hands on.

An Important Note About DAWs

I try very hard to be software-agnostic when it comes to what audio software you use. That means that this book can be used with any digital audio workstation (DAW) as long as it has the necessary processors and plug-ins to create a great mix. 99% of all audio software today has the necessary plug-ins you need, but there is some software out there that are more audio "editors" than audio "workstations" – meaning that they might not have the most flexibility when it comes to mixing.

Personally, I use Logic Pro X and I love it. However, I've also used Pro-Tools in the past and at the Icelandic Embassy Studios we use a combination of Cubase and Logic Pro X because neither me nor my business partner wants to learn the other person's DAW.

It's also important to stress that no DAW is better than another. It's really all depends on the skills of the user. I won't switch back to Pro-Tools because I've become so familiar with Logic throughout the years that it simply makes no sense for me to switch back. A lot of musicians think that having Pro-Tools is what makes a recording studio professional. They're wrong. A recording studio is professional when it has knowledgeable audio engineers that take care of their customers and make them comfortable enough to record a great performance. Once you've got a great recording on disk it's not the software you're using that makes your mix good, it's your knowledge of mixing and your skills that make for a great mix.

Finally, it's up to you to learn how to use your DAW. The technical details of how certain things work in the DAW of your choice might be slightly different. For instance, using groups and busses is a standard way of organizing your sessions, but every DAW does it slightly different. So if you don't know how to use your DAW to organize tracks, import files, use the arrange window or the mixer, you don't need to read this book yet. You should start with your DAW's manual! In this book I will be using general terms you can use in any DAW. However, it is up to you to know how to do them inside your DAW of choice.

STEP BY STEP
MIXING

HOW TO CREATE GREAT MIXES USING
ONLY 5 PLUG-INS

Step-By-Step Mixing • How to Create Great Mixes Using Only 5 Plug-ins

Chapter 1 – Pre-Mixing

Even before you start mixing your song, you need to make sure it's ready for mixing.

Some people produce their songs and mix them at the same time. That's a great method if you're acting as the producer, recording engineer, and mixing engineer all at once. That means you have a vision for the song from the start and know exactly where you want to take the production throughout the entire process.

WHEN DOES MIXING START?

So let's talk a little about what it means to "start mixing."

For our purposes, we'll be looking at mixing as a separate phase of the production process. We'll be imagining that somebody else did the production and the tracking, and somebody else will handle the mastering. So your job is simple: pull up the faders and get to work.

But if you're involved in the entire project from start to finish, the lines get blurred a little bit. If you're tracking the project and deciding on the sounds and production of each instrument, you're kind of adding a bit of your mixing mentality to the entire project. You have an idea of how things ultimately should sound so you basically do everything with the end in mind.

Either way is cool, there's just an interesting mental shift in your thought process depending on when you get involved in the process. You'll end up doing a lot of the same things to the mix regardless of how much you're involved, but you're still coming at it from two different points of view.

Now, either way, you can't start slapping plug-ins and processors on until the recording, production, and editing process is over.

Let's imagine that we've finished recording all the instruments and all the overdubs are done. That means that you have a finished song with all the tracks, ready for the next step.

And that step is... not mixing.

Before you can mix, you need to get your recorded tracks ready for mixing. You need to clean up all the crap in your tracks.

DO THE DISHES SO YOU CAN COOK IN THE KITCHEN

Here's an analogy for you, let's see if you can keep up.

Cooking is great. Doing the dishes afterwards is terrible. And it's even worse when you leave them in the sink until the next time you have to cook. Then you not only have to do the dishes before you cook, but you ALSO have to do them afterwards!

Terrible right?

It's the same with mixing. If you clean up your tracks before you start mixing you'll have a more enjoyable experience. You don't want noise from the vocal tracks creeping in during the solo.

So before you start mixing, do these two things:

- Trim the regions and delete the "noisy" silences between parts.
- Add fades to all the regions so they don't abruptly pop in.

Trimmed regions with fades to eliminate noise in between instrument parts

Going back to the "doing the dishes" analogy, if the kitchen is clean I know where everything is and I can grab it immediately. If I leave the dirty dishes in the sink, I'll inevitably have to scour through the dirty dishes to find that one utensil I specifically need for my cooking.

If you don't edit your tracks first, you'll run into similar problems. You'll notice an annoying click somewhere, a misaligned drum hit, or background noise you should have edited out. This hinders your workflow because you're constantly going back and forth between the mixing and the editing phase.

If you're always changing hats then you'll never fully focus on one aspect of your production.

It's better to cook with a clean kitchen. Having a full sink of plates and utensils you might need is just going to rob you of the pleasure of making my meal. The same goes for editing. You'll rob yourself of the joys of mixing if the editing phase is constantly nagging at you in the back of your mind.

And like dishwashing, editing is pretty boring. It's tedious and usually pretty uncreative. But if you gloss over it and ignore it, you'll be left with a sub-par production.

So as a general rule, edit first, then focus on the mixing. Once you're done editing and everything sounds and feels great, then you're ready to start mixing.

Chapter 2 –
Mixing With NO Plug-ins

Before we get started with using EQ, Compression, Saturation, Delay, and Reverb, let's talk about something that's very
important to do even before you start slapping plug-ins on your
tracks.

ORGANIZE!

The first thing you should do before you start playing with your
plug-ins is to organize everything to make your mixing process
faster.

Before you even start pushing faders around, make sure you organize your tracks so you can fly around the session, knowing
exactly what you're looking for.

A good way of doing this is to use color-coding. Having all your
tracks and regions assigned a different color based on what instrument or group of instruments they are helps your workflow.
Different colored tracks help your eyes and brain jump right to
the correct track when you scroll from one side of the mixer to
the other.

This gets especially handy when you've added a bunch of groups and you have an even larger session than what you started with. Also, it's a good idea to stick to the same color scheme when you're mixing so that you train your brain to know exactly what each color is in every session.

For instance:

- Drums and percussion are Red
- Bass is Purple
- Acoustic Guitars are Orange
- Electric Guitars are Green
- Vocals are Dark Blue
- Keys, random pads and such are Purple

Of course, this is just an example.

Example of color-coded tracks that are easy to find. (screenshot from the Mixing With 5 Plug-ins video course you can find at www.stepbystepmixing.com/resources)

You're allowed to use your own color preferences, of course! You can go even further with this scheme and get more elaborate. In fact, the reason my electric guitars are green is because Andrew Scheps (mixing engineer for RHCP, Beyoncé, Adele and U2 to name a few) talked about it at an AES conference a few years ago. He mentioned that he even goes so far as to relate the gradient of the green to the cleanliness of the guitar parts. So the cleaner the guitar tone, the lighter the green.

So knock yourself out with color-coding, but don't get so wrapped up in it that you don't continue mixing!

Now that we've color-coded we can start actually balancing the tracks.

CRITICAL LISTENING AND VISUALIZING YOUR MIX

There are multiple ways to start mixing and start balancing the tracks together. A great way to begin is to simply listen to the song and listen to what you hear with all faders up. That way you notice immediately if anything is too loud or too quiet.

Look at all the tracks you're working with and think about what your end goal should be. Is it a big EDM track where you have to sift through countless tracks of samples and rhythm before you can actually get going on your mix? Or is it a sparse track with guitar and vocals and percussion that all need to be big in the mix.

The type of production you're working on will influence your mixing decisions. The big 100-track EDM song will require a lot more sculpting with EQ than the sparse, 4-track acoustic song that can fill up the mix easily without any of the instruments crowding the mix.

The same thing will apply to your use of compression and effects. Big pumping compression effects might be exactly what you'd like to hear in an EDM mix... but the acoustic guitar won't sound so good when it's pumping really hard. It'll just sound out of place.

As you're listening you can also watch the waveforms in the arrangement window so you can see the overall loudness of the tracks. Doing this helps you notice tracks that might be very important but are too low in level compared to the other tracks. You don't want those to slip through the cracks because they might be really cool parts of the song! For instance, a lead guitar track might be a really important part of the song, but if it was recorded too quiet compared to everything else it will get buried within the other tracks.

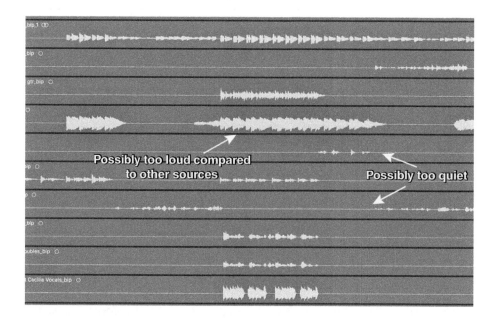

You can't expect all tracks to magically balance themselves so make sure you jot down any things that stand out throughout the initial listen to your mix. Maybe the drums are pretty balanced on their own but the tambourine is a bit loud, or the bass is pretty low in the mix and all the guitars are kind of one big mush. On top of that, the vocals might need some focus on the lead singer and not as much on the backup vocalist.

Keep a critical ear on everything that's happening in the tracks and pay attention to how each track relates to the overall mix. However, and this is why it's so important to listen with a critical ear when you approach a song you're about to mix, you need to make sure that the best parts of each instrument are highlighted.

It's entirely possible to make a really great sounding flat mix where everything is audible but nothing stands out. This makes for a great rough mix, but might sound boring and two-dimensional because there's nothing that really drives the music and makes it exciting.

This is where automation and pulling things up and down in the mix come in to play. It's not just in the parts where the guitar and keyboards actually groove together and are the main focus (which is important to pull up), but it's also about listening to each track and spotting interesting phrases in the performance that you can lift up and draw attention to.

You might choose to leave some of the nuanced automation stuff until the final mixing phase, but it's good to make notes about it during your initial rough mix.

As you're listening to the song, do this exercise I like to call: Think tall, wide, and deep.

■ TALL – FREQUENCIES

With your eyes closed, what needs to be tall? Which elements need high frequencies and which need low frequencies?

Which instruments can you trim the top off of and place in the low end? Which instruments can you filter the fat out of and make shine in the high end? What needs to be all-encompassing in the middle?

▨ WIDE – PANNING

Think about where each instrument should sit from left to right. You need to fill out the stereo spectrum, so pan your instruments around your canvas. Artists are supposed to fill out their canvases, not just paint in the middle.

The same goes for you. Don't place everything square in the middle of the stereo spectrum; instead, spread it out and find a good spot for each individual instrument.

▨ DEEP – REVERB, DELAY AND EFFECTS

Think about which instruments need to go back and which you should pull forward. Mentally visualize moving them closer or pushing them further away. Think of the mix like a 3D image.

With volume, reverb, and effects you can place anything anywhere in the room. For example, if you want the drums by the rear wall, you know that you'll need to add some reverb and lower the volume of the dry drums. You push back with reverb and effects; pull forward with more volume and short delays.

Once you've visualized all the elements in terms of highs and lows, left and right, and front and back, you have a better idea of where you're headed and what you need to do.

This exercise in critical listening is something you can do without using any plug-ins. So make it a habit for all of your future mixing sessions: Just listen to the song all the way through and take notes on where you want everything to go in the frequency spectrum, the stereo image, and the depth of field.

INITIAL BALANCING AND HEADROOM

When it comes to balancing, I like starting with my levels fairly low because I know that they'll get louder as I go through the mix process. This makes you really focus on hearing what's going on in the mix and what effect (if any) some of your mix moves are having.

For instance, it's especially great when you're mixing the drum bus. It forces you to really listen to how much difference just a few dB changes here and there with an EQ can make. If you strain to hear the changes you make at low levels it becomes easier to mix because you're training your ears to be more sensitive. It's a great exercise in focus and attention to detail.

When I do an initial balance I like to move all the faders down a few dB under unity gain so there's more headroom to work with. I've found that moving the tracks down so that they all peak at about -11 dB is a good starting point. If certain tracks are too quiet or loud I can adjust the volume of each track without overloading the channels themselves and clipping the faders. You don't want to be pushing the fader of one track all the way up when you can just lower the louder channels in comparison.

From there you simply massage the faders in place. Don't worry about the balance being absolutely perfect right away. You might find that certain instruments are too loud in the chorus compared to the verses, or vice versa. Or maybe the vocal is very dynamic and you just can't seem to find the right spot for it. Don't worry too much about that during the initial balance.

Balance is just the first step. We'll be adding EQ to create spaces in the frequency spectrum and compression to take care of dynamics soon. However, spending a little extra time finding the best balance to begin with will save some time in the long run.

Also, try not to isolate tracks and solo anything too much in the initial mixing stage. Instead, push the fader way up to hear what it's contributing to the mix. Then move it back down and blend it into the context of the track. This is very important and can be a better way of hearing things as they relate to the rest of the track. For instance, if I have multiple kick tracks I just audition the kicks in context of the mix, then I'll push them down until they fit with the drums and move on. Sometimes it's more about speed than pinpoint precision.

The same goes with overheads, multiple rhythm tracks, or a group of backing vocals. You just push the fader up, hear what it's doing, then move it back down until you feel the balance is right.

MIXING IN MONO

Mixing in mono is an important part of the mixing process. Personally, I start my mix and mix almost entirely in mono on a one-driver Behringer Mixcube for the first few hours until I have a rough mix in place. If you can make your mix sound good in mono on a Mixcube then you'll be pleasantly surprised at how good it sounds when you finally flip it into stereo and push it through your better sounding studio monitors.

In short, if you can make your mix sound awesome in mono on a Mixcube it'll usually sound awesome everywhere else.

Next time you're mixing, try to mix at lower levels than usual, in mono, on a really crappy speaker. It's kind of like athletes that run with weights on their feet. They make it deliberately hard to train and give themselves an uphill struggle to begin with so that when the weights are off everything feels easier.

PANNING IN MONO

Panning is important. You don't want all of your tracks fighting for the center. For instance, panning the drum-kit is an important way to expand and establish the stereo spectrum.

If you have all sorts of different elements and instruments, you need to find a place for them in the stereo spectrum. Pan everything around until you've found a good balance. Whether you're a hardcore Left-Center-Right panner or like to spread things around in increments around the pan knob doesn't matter. Keep in mind that you don't want to tip the balance of instruments too much to the right or left either. Instead, try to find a good equilibrium and balance between the left and right speaker.

A good way to do this is to keep things in mono and pan that way. You'll find it easier to create separation in your instruments if you don't have to listen to your pans in stereo. Counter-intuitive? Maybe. But it works.

Panning in mono gives you a different way of placing the instruments in your mix. You're not panning in the stereo field anymore since you've flipped your mix to mono, but you can still hear a difference in the separation of instruments by doing it this way. I'm not entirely sure of the technical implications of this, but my guess is that it could have something to do with the relative level of the panned tracks compared to the centered tracks. That's about as plausible explanation I can think of without knowing the exact physics behind it.

Then, once you flip your mix to stereo you'll get that awesome feeling of your mix completely opening up wide. You might not find this method mentioned anywhere else, but I've found it very helpful when approaching my mixes. It's ok to experiment, even if you don't know why something works!

BALANCING THE DRUMS

Let's talk about balancing the drums. Since the drum-kit is a combination of several different instruments and microphones it can give you multiple options (or headaches!) depending on what kind of sound you're looking for. If I'm looking to go with a more natural drum sound, I tend to emphasize the overheads and keep the room mics more present than the individual drum tracks. You can almost look at a room mic as just a reverb mic that makes the drums larger or smaller depending on how loud they are. Conversely, if you want a "tight kick and snare" kind of mix it's a matter of pulling the overheads down and letting the kick and snare close mics do most of the talking.

You can get even more detailed if you have multiple tracks of each individual drum, like if you have an over and an under-snare microphone. Want a snare sound that's heavy on the snare rattles? Then just make the under-snare microphone louder. Want more attack? Accent the top mic.

The overheads also affect the width of your mix. If you want a wide mix you should pan your overheads hard left and right. If you want to keep the drum mix more centered, a narrower overhead pan will help you do that. That's really more of a taste issue than a hard rule, so feel free to experiment as you like. I've heard all sorts of different drum sounds, from open and wide to completely panned to the right (hi there Beatles!).

One of the things I often struggle with is the level of the toms. You want them to really ring through whenever the drummer does a fill, but you don't want them to sound detached from the rest of the drum sound. The right combination of overhead sound and individual tom balance is important to achieve this, but if you're not focusing on the overheads that much it might be necessary to add depth to the toms with reverb, as well as keeping them consistent in the drum mix with compression.

When you've got a good drum balance in the context of the rest of the instruments you can move on to make the rest of the tracks fit. The bass can be pushed up to where it sits audibly in the mix without adding too much clutter in the low-end. Don't worry if the low-end sounds too boomy or muddy. You'll use EQ to tame the low-end later - there's always a need for rebalancing throughout the entire mixing process.

Once you move onto guitars you need to think about what role they are playing in the mix. Is it a predominantly guitar-driven song or are the guitars just for backup? That will influence your level decision. Is there a combination of multiple guitar parts playing different things, or a combination of both acoustic and electric guitars? That will not only influence the balance of the mix once you decide which guitars are more important, but it also influences your panning decisions as you don't want them to clash in the stereo spectrum.

Also, are the guitar tracks doubled? Do the doubles play all the time or do they come in during various parts of the arrangement, like the chorus, to lift the song up to another level? If that's the case you need to make sure it doesn't accidentally skew the mix to either side of the stereo spectrum or drown out the regular guitar parts.

In a predominantly guitar-driven song you also need to think about your guitar levels relative to the vocals. You don't want to drown out the vocal because it's the most important part of the song, and you also don't want the guitar crowding out the vocals in the center of the mix. You might get a lopsided mix if there's only one guitar part and you pan the guitar out of the way. If that's the case you have the option to give them both space in the center with EQ and/or reverb. But if you can, panning the guitars out of the center is a good way to make room for the vocal. Alternatively, if you only have one guitar part but a doubled vocal, panning each vocal track -10 and +10 to the sides will move them slightly out of the center and free up some room for the guitar.

But enough about guitar balancing. I could go on for hours here.

THE 1 DB RULE

The 1 dB rule is a trick I stole from my friend Mike Senior, who wrote about it in his book "Mixing Secrets for the Small Studio." It came from one of the mixing engineers he interviewed in the book, I forget whom, but it's a rule I really live by when I'm struggling to get an initial static mix.

Here's how it works: move each track up or down 1 dB. If the mix sounds better try another dB until it starts sounding worse. When you can't move a track one dB up or down without making it worse, you're as close to the sweet spot as you can get with only levels. The same goes with panning. Just moving certain instruments away from each other in the stereo spectrum cleans up the mix.

You really follow the same pattern with other instruments, whether it's drum loops, keys, extra acoustic instruments, synthesizers or vocals. Massaging them in place and moving each track up or down 1 dB at a time will help you get a nice initial balance.

This is a technique I demonstrate in the Quick Mixing, Mastering and Editing training you can get inside the Step By Step Mixing resources: www.StepByStepMixing.com/Resources.

DYNAMIC PROBLEMS DUE TO ARRANGEMENT

Now, you might be thinking, "well, it's great to get a good static mix in one part of the song, but the choruses are much louder than the verses so my balance is all out of whack as soon as the chorus hits!"

This is a very common problem! We won't be fully fixing it at the start because the best way to do so is with automation and we'll cover that later, but here's what I recommend:

Create a static rough balance of your mix at the loudest part of the song.

The loudest part of the song is very often the last chorus because that's where all the overdubs, doubles, and extra vocal parts are found. It's the hardest part of the song to balance because of the number of tracks you need to balance together. But if you can get a good static mix there, the rest of the song will fall into place more easily.

THINK FAST, NOT FLAWLESS

There is no one way to get a good initial balance, and there's certainly no rule about where you should pan the instruments. Nobody's lost a mixing job because they panned the shaker too much to the left.

The goal for the rough mix is to create as much separation between the instruments in the stereo field (even if you're panning in mono) and create a balance among all the different instruments with volume.

What kind of balance you're looking for is up to you. If you're a guitar player chances are you'll want your guitars loud. If you're a self-conscious vocalist like me then you'll probably make your vocals quieter than if you gave the song to a different mixer.

Don't worry too much about getting a perfect balance to start with. You want to move fast. There's a reason we call it a rough mix. It's rough because you haven't added any processing to make those instruments jump out. In Mixing With 5 Plug-ins video course, I manage to get a rough mix going in about seven minutes, but that might be a bit too fast for you. I would recommend aiming for about 20 minutes at first to get a rough mix going. Any longer than that and you'll be doing micro tweaks that aren't necessary at this stage of the mixing process.

GROUPS

If you get a good static mix with all the individual tracks and you're happy with the balance, you can make the rest of the mixing process a lot easier by grouping tracks together into subgroups and busses. This helps you narrow down the entire mix into just a few different elements. That way you can use plug-ins on all those busses instead of on each individual track. It saves processing power and helps you mix faster.

Here's a common grouping you could use for any mixing session:

- **Drums and Drum Loops** – Drums go into one subgroup together. You could also make this more complicated by routing all of the kick drum mics into one kick group that's then routed into the drum group. You can do the same if you have multiple snare tracks and also for the toms. If you're working with virtual drums, the process is the same. If you're working in a genre where you're using simple stereo drum loops, you already have the drums grouped. However, you may want to group them together with any other percussion instruments if it makes sense for you to simplify in that way.

- **Percussion** – If you have multiple types of percussion it could be in the same group as the drums, or in a separate group by itself.

- **Bass Tracks** – Bass should be a separate group, whether it's a bass guitar or a synthesized bass instrument. You might not even need to route it to a group, but for my workflow it keeps me organized just because of the way Logic Pro X works with groups.

- **Guitars** – Depending on the types of guitar tracks you have there are multiple options for groups. If you just have a bunch of distorted rhythm tracks then they can all be routed to one group. But if you have multiple rhythm guitar tracks, acoustic guitar tracks, and multi-miked solos, then you'll want to keep them fairly separated. It really just depends on the arrangement how much you can simplify your mixing through groups.

- **Vocals** – With vocals I tend to have Lead Vocals as a group, and then a Backup Vocal group if there are lots of harmonies.

You may be working with completely different instruments, but the main goal here is to recognize what tracks naturally go together so that you can simplify your mix and make the rest of the mixing process easier.

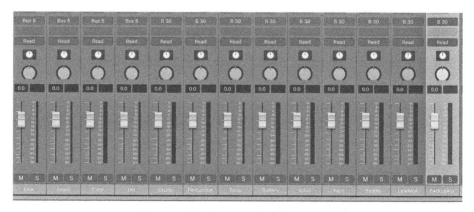

Example go-to groups from my mix template

That's a simple starting point, and you could go even further and route all the instruments into an All Instruments group for further processing. However, if the goal is to simplify then you don't want to end up with more group tracks than there are tracks in the session to begin with!

Another reason for groups, other than simplification, is that they let you easily rebalance and adjust each generalized component of the mix. If you're happy with your drum balance but just want the drums a little louder overall, it's easier to pull up one group fader rather than moving every single fader on the individual tracks. Groups can even help you visualize how different elements are combined together, which will also come in handy when you start adding effects.

THE 3 WAYS TO APPROACH AN INITIAL MIX

If you've done what I've talked about so far, your session should look organized and color-coded, with each mix "element" routed to a different bus.

The great thing about busses is that they allow you to conserve processing power by using a single plug-in on multiple tracks. By doing this you also have to decide where you're going to start using plug-ins, which brings me to the three different ways of approaching the initial mix.

1. Individual Track Mixing

   If you choose not to do any channel grouping and would like to mix each individual track separately then this is what you'll end up doing. Maybe the song isn't that complicated and adding busses doesn't really save you any time or processing power. If that's the case just slap a plug-in on every track and mix the tracks together that way.

2. Top-Down Mixing

   Top-down mixing is when you start mixing your song with plug-ins on the master bus. After you've got a good static mix going, you'll start by adding EQ, compression, and other plug-ins that give your mix color (such as analog summing plug-ins or mild tape saturation) to the mix bus. It's actually a great way of starting your mix because you'll immediately get some big wins from tightening up your mix with compression, tweaking the overall EQ response, and adding some secret sauce with saturation.

   From there you move from the top (master bus), to the group busses, and finally end up at the individual tracks.

With this method you might actually find that you need to do very little processing at the individual track stage because you've already made some big strides with your mix by applying processing to the master bus and the subgroups.

3. Middle-Out Mixing

Middle-out mixing is when you start using plug-ins on the subgroups instead of the individual tracks or the master bus. (I don't know if "middle-out" is an actual term or not, but it's what I call it.) I tend to do a combination of middle-out and top-down when I mix. Sometimes I just put a compressor on the master bus and then start focusing on the subgroups. If the subgroups sound good but there is an overall frequency issue in the entire mix, I like to try fixing it with a master EQ instead of drilling down into the tracks themselves.

Once I tweak the subgroups and the master bus I'll turn to the individual tracks if necessary. Usually the individual tracks can benefit from slight EQ tweaks, especially if you're trying to separate things in the mix, like two similar sounding electric guitars. Other times you need to add compression to reel things in, or make them hit harder. The kick and snare come to mind in that scenario.

RECAP

Alright, let's wrap up this chapter with a quick recap.

If you've been following along (whether that's in your head or inside your DAW) here's what you should do before moving on to EQ:

- Arrange your tracks so you know where everything is.
- Keep the same lineup of tracks in your sessions so that it's easier to find everything.
- Color code your tracks to train your brain to jump to the right channels when you're working.
- Spend time listening to the tracks and the song and jot down what jumps out at you that needs special attention, like very dynamic vocal tracks or softly recorded guitar tracks that drown in the mix. Use critical listening and jot down all the things that feel exciting about the song.
- Sometimes you have tracks that will need different processing for each part of the song. For instance, a vocal might need different EQ and reverb in the chorus than the verse. If that's the case, see if you can "mult" (split the audio track into two different tracks), one for "Verse Vocal" and one for "Chorus Vocal."
- Spend time on setting levels. Push up each track a little bit here and there and notice the subtle differences it makes. If the levels are all over the place inside each track then it might be hard to find a level that's consistent. You might have to fix that with compression, or by using automation right away.
- Finally, simplify your mix by creating subgroups and using busses for an easier way to visualize each element of the mix.

Chapter 3 –
Using EQ

Now it's time to start using EQ. In mixing, EQ is basically used in two different ways - to clean up some of our tracks and separate them from each other so that each instrument has its own frequency space in the mix.

You should be prepared to use EQ on every track, starting with the busses and then move to the individual tracks.

Note: In this chapter I will give you all the necessary knowledge to get started using EQ in your mixes, whatever genre you're working in. However, EQ is one of my absolute favorite subjects so I have even more information inside EQ Strategies - Your Ultimate Guide to EQ. You can find more information about that inside the Step By Step Mixing resources: wwww.StepByStep-Mixing.com/Resources.

WHAT EQ SHOULD YOU USE?

We can debate which EQ is the best all day long and we'd both be wrong. The same goes with most stock plug-ins and processors. Most software EQs are very comparable. They usually all do the trick. Some plug-ins might have slightly different modeling, but for our purposes a simple visual EQ will work just fine. Usually the stock EQ is the only thing you need. If you have some sort of "vintage" EQ plug-in that sounds slightly different you can go back and forth between them depending on what you're working on.

For instance, I love the Fabfilter Pro-Q 2 EQ plug-in. It's extremely versatile and it gives you a bunch of cool options for finding the right frequencies you're looking for. The actual EQ itself is probably just the same as Logic's stock EQ, but the interface is more intuitive and easier to use. I also like using it for tutorial purposes because it's so easy for students to follow along on the screen.

In addition to my Pro-Q 2 I tend to use specific EQs that emulate some sort of hardware, like the Waves V-EQ. Although it has a little less flexibility, I tend to use those EQs for color later on in the mix, or to add something back in that might have been lost due to compression.

THE 3 PRIMARY FUNCTIONS OF EQ

There are two types of EQ: Graphic and Parametric. Most of your channel EQ processing will benefit from the parametric EQ because of its precision.

Any parametric EQ has three parameters:

- **Frequency** – Where you select which frequency you want to manipulate.
- **Gain** – Where you decide whether you want to increase

(boost) or attenuate (cut) the frequency you've selected.

- **Q** – Where you decide how much you want to affect the frequencies around the one you chose. This technically defines how wide or how narrow the selected frequency bandwidth will be.

You'll find these three things in every software EQ in some way. Plug-ins that emulate hardware might have fixed frequencies and Qs, but the overall function is always the same.

THE 3 BASIC METHODS OF EQ'ING

Although an EQ might look pretty intimidating it's actually fairly simple to use. There are really only three things you can do:

- **Filter** – When you filter frequencies you eliminate them completely. High-pass filters are frequently used to eliminate unnecessary low-end frequencies from a mix and low-pass filters are common to tame the high-end hiss.

- **Cut** – Commonly referred to as subtractive EQ. When you cut a frequency you reduce its power in the frequency spectrum. Sometimes you need to cut annoying ringing sounds and resonances in a specific track. Sometimes you need to cut certain frequencies to make room for other instruments in a mix.

- **Boost** – When you boost frequencies you're adding more of them to the mix. You boost to add something that's missing, like presence in a vocal track, attack to a guitar, or body to your bass.

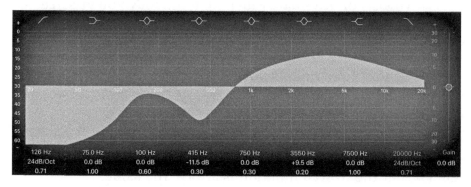

This EQ has a high-pass filter, a low-mid cut and a high-mid boost.

Now that you know your way around an EQ plug-in, let me give you some great guidelines that work in almost every mix.

■ FILTERING USING HIGH AND LOW-PASS FILTERS

Filtering is the most basic, and to a certain extent the most destructive, part of EQ'ing. Filtering is basically when you eliminate all the frequencies either above or below a certain cut-off frequency. It's like cleaning up the clutter before you can make your room nice. All instruments have frequency ranges that get in the way of other instruments in a mix. However, don't fear the filter. It's the best way to eliminate low-end buildup and clutter from instruments that don't need it.

For instance, when you apply a high-pass filter to a vocal up to 100 Hz, you're letting the "highs pass through" unaffected above 100 Hz. Conversely, if you use a low-pass filter on a kick drum down to 10 kHz then you're letting all the frequencies lower than 10 kHz "pass through" unaffected. It's a bit backwards, but just remember what frequency range you're letting "pass through" and you should get the hang of it.

You'll notice that you can select different slopes on certain EQ filters. Sometimes I use different slopes and that's partly because the gentler (more gradual) slopes are supposed to be more musical and more natural sounding, so I use them on instruments I'm a little more heavy handed with (or filter more, that is).

I want you to have an easy reference tip sheet (like the one in the Step By Step Mixing resources at www.StepByStepMixing.com/Resources) every time you need some helpful tips on EQ, so I'll break the EQ section down into filtering, boosting, and cutting tips so that you can easily find the tips you're looking for each time.

The following tips are good starting points and recommendations that I've used to build my mixes. As always, experiment with what sounds best for your tracks. And remember – getting a good sounding recording at the beginning with your mics and pre-amps will go a long way in helping you get a great sounding mix once you start using these techniques.

DRUMS

- Kick drums are bass instruments so they're rarely filtered above 32 Hz.
- You can reduce bleed from the rest of the drums by low-pass filtering the kick drum down to the high-mids, about 5-10 kHz depending on what sounds good.
- You can take some of the oomph out of the snare drum (especially if the kick is bleeding into the snare mic) by filtering everything below 100 Hz.
- Depending on what kind of style you're going for, you can filter quite a bit of low-end out of the overhead mi-

crophones. If you want a natural overhead sound, or if the overheads make up most of the drum-sound anyway, then filter as little as you can. If your kick and snare tracks are supplying most of the sound and you're just looking to accent the cymbals and add some ambience around the close-miked kick and snare, then you could filter up as high as 500 Hz.

- If you're working with virtual drums, the same rules apply. However, if you're using drum loops that include a sampled kick drum, you might want to go easy on the filtering so that you don't lose the low-end "weight".

■ BASS

- I rarely high-pass filter the bass higher than about 40 Hz. Usually it's just to get rid of any low-end rumble it might have.

- If you're looking for a rounder bass that functions more like a pad than a string instrument then you can filter out all the highs to just accent the lows and the low-mids. Just move the filter down into the mids until it sounds right.

- Sometimes there's really nothing going on above a certain frequency. You'll be able to see the frequency representation if your EQ has a frequency analyzer, so you'll often be able to gauge how much you can filter by looking at the analyzer. It's a good crutch, but don't rely on your eyes to mix!

GUITAR

- A good starting point is to high-pass filter your guitars up to 100 Hz.

- You can high-pass your guitars even higher, depending on the arrangement and how busy the mix is. If your guitars are clashing with other instruments in the low-mids, chances are you can fit them together with the right filter frequency (or a parametric boost).

- If you're worried about filtering too much, just solo your guitars and sweep the filter up the frequency spectrum until the guitar (or any other instrument) starts to sound too thin, then back off to just before that.

- It might sound great in solo but you may want the guitar to sound a little thinner to fit in the mix, so make sure you A/B the filtered guitar in context with the rest of the mix.

- One of my favorite filter tips for distorted electric guitars is to use a low-pass filter that combines a boost at the cut-off frequency. Most filters will let you add a resonant boost at the cut-off frequency if you increase the Q.

KEYS AND PIANOS

- Depending on the performance and what register the keys are playing, I would advise against filtering too much.

- A good bet for most instruments that aren't "bass" instruments is to filter up to 100 Hz, but if you feel there's something lacking from the instrument at that point, you can move your cut-off frequency lower.

- Try my "resonant filter + boost" tip from the guitar section on rock organs or hard-hitting keyboards that don't need (or have) a lot of high frequency energy.

■ VOCALS

- Depending on the gender of the singer, you can often get away with filtering out quite a bit of low-end energy. Start at 100 Hz and move up until you feel the voice thinning out.
- If you want to blend the backing vocals and keep them fairly dark in the mix, a low-pass filter can help with that.
- Alternatively, if you high-pass filter the vocals quite heavily and increase the volume you'll get them to stand out a bit more, giving them more of an "airy" quality that cuts through the mix.

■ OTHER INSTRUMENTS

- Regardless of what genre you're working in, it's a good bet that your tracks can benefit from filtering either the highs or the lows.
- If your instrument is sounding too thick in the low-end, add a filter to clean up the bass.
- If your instrument has a lot of high-end energy that sounds shrill, hissy, or fatiguing, adding a low-pass filter will clean up the unnecessary air.

Filters are a fairly simple affair, but you'd be amazed at how much cleaner your mixes sound once you've added a few strategic high-pass filters to get rid of the low rumble and the low-mid mud. Adding a few low-pass filters will also clean up any hiss from hard-rock guitars, get rid of unnecessary bleed from drums, and blend backing vocals. They should be your first tool to get your instruments to fit together better.

However, don't think of filters as their own separate thing. They usually go hand in hand with whatever other EQ'ing you're doing, so let's talk about how you can attack problematic frequencies and accent the frequencies that make your instruments cut through the mix.

SUBTRACTIVE EQ

When you're dealing with home recordings, chances are you'll have weird resonances from your room in your tracks. It'll be a frequency or two that somehow gets magnified way too much and it'll make your tracks ring in weird ways. This can come across as boomy bass, harsh guitars, or annoying ringing in your snare sound. That's when subtractive EQ comes in.

When you cut instead of boost, you're taking away frequencies from your tracks. It helps you repair less-than-ideal recordings. It allows you to surgically remove problematic frequencies so that the tracks sit better in the mix.

My quick and dirty method for subtractive EQ'ing is really simple. I sweep around the spectrum with large boosts until something ugly pops out, then I cut it until it sounds smoother.

ADDITIVE EQ

EQ boosts are by far more fun to do. Subtractive EQ is incredibly important but it's not always as fun because you don't get as much out of cutting frequencies like you do with boosting. Maybe it's just the way our ears work, but it's simply more gratifying to add more frequencies (such as high-mids) to make instruments cut through the mix.

Unfortunately, that's also where the danger lies. Adding too many boosts can cause phase issues, although that shouldn't discourage you when you absolutely need to use EQ boosts. The main danger is that once you start adding high-mids to one track for instance, you'll be rewarded with how much better that track sounds in the mix. That can lead down a slippery slope of adding high-mids to everything until you end up with a really harsh and piercing mix instead of one that's balanced in the frequency spectrum.

Just make sure you're aiming for balance in your mix, where all of your instruments can be heard well in the frequency ranges where they sound the best.

To help you with that, here are some tips to keep in mind as you EQ your mixes.

Please note: Every instrument and track is different. As such, the exact frequency where you'll cut or boost can vary. Frequency charts can be helpful, but relying on the numbers without listening to the music won't lead to better mixes.

The following tips are guidelines to start with, but make sure you spend time finding the exact frequencies that make your mixes sound better. I'm focusing on the "big wins" that will help you improve your mix the fastest by giving you solutions to the most common problems I've encountered throughout my years of mixing.

▨ DRUMS

- Cut the kick drum in the 300 - 600 Hz region to get rid of boxiness. Sweep around with a narrow Q and a big boost to find the boxiest frequency and then get rid of it.

- The same works well on a drum group bus, toms, or any other drum that needs to be smoothed out. Start with a cut around 400 Hz and see if your drums tighten up a bit.

- If you have multiple microphones on drums, such as an over and under snare mic, then make sure you check the phase and polarity relationship between all the drums. You'd be surprised just how much punch you can add back into a drum sound if you just make sure everything has the same polarity. This tip also applies to any instrument that's multi-miked. Flip the polarity of one of the tracks and see if it doesn't add some extra weight and power to the overall sound of the instrument.

- The typical philosophy for kick drums is to cut the mids and boost the lows for bass and the high-mids for the beater. Although I've found this to be true most of the time, some genres need a heavier hand to tame the lows. Metal kick drums can get muddy really fast if there's too much low-end. You might want to add a shelving cut filter to tame the lows while you add a big boost to the beater area around 4 kHz. Let the bass guitar handle the low-end presence and make the kick cut through in the mids instead.

- One general rule of thumb I follow is: The harder the genre, the more higher-mid boost I use. Metal kick drums come to life with a big boost in the 4 kHz area, but for softer genres like pop, folk, and rock you can get the results you want by focusing more on the area from 1.2

kHz to 3 kHz as a guideline.

- Harsh cymbal noises can be tamed with a cut in the 2.5 kHz area. When you hear the cymbals piercing through the mix don't assume you need to cut the highs because that's not the part of the cymbals that's being annoying. Usually the high-mids cause the most problems so you can still get a clean and airy cymbal sound without the annoying harshness.

- Home recorded drums often have annoying low-mid buildup, causing the kick drum to have too much energy in the 100 - 250 Hz area. Don't be afraid to cut lows and low-mids in order to clean up your drum sound. It doesn't always take a low-frequency boost to create powerful bass. Sometimes it's about cleaning up the area to hear the bass that's already there.

- However, if you need more low-end oomph in your kick drum then find the frequency that sounds good to you by boosting around 60 - 100 Hz.

- If you want to add body to your drum sound try hunting for it around 150 - 250 Hz. If you like a meaty snare sound then boost the low-mids to bring out the thickness of the body.

- Alternatively, if you're looking for more sizzle or attack, bringing out the 2.5 - 3 kHz area can help bring it out in the mix.

- If that brings out the rattle of the snares too much, a high-shelving boost around 10 kHz will increase the brightness of the snare drum without adding harshness from the snares themselves.

- The area around 2.5 kHz is a good starting point to bring

out the attack of any drum, whether it's the snare, toms, or the beater of the kick drum. But just like I said before, it also causes harshness if boosted too much.

EQ'ING DRUM SAMPLES AND LOOPS

Since drum loops and samples are just pre-recorded drums you can use the same EQ guidelines to sculpt the loops however you want. The same principles apply whether you're working with a rock kit or an electronic loop.

The only thing that's different is how you want the loops to sound in the genre you're producing. The lows are still the lows, the boxiness is still present in the same place, and the high-mids will still hold all the attack, tone, and harshness as before.

I'd suggest that it's easier to EQ loops and samples because they've already been pre-produced and mixed so all you need to do is to EQ them to fit with the rest of the arrangement instead of trying to solve any particular frequency problems, home record-ed resonances, or pesky problematic frequencies.

▓ BASS

- A lot of the same rules apply in the low-end for bass guitar and the kick drum. They like to clash and fight in the low range, so make sure you give the kick drum its space in one specific frequency while giving the bass some room to breathe in another.
- Soloing the kick drum and bass guitar while allocating the right frequencies to each of them can be helpful.

Sometimes the kick might sound slightly thin, but once the bass is added to the mix it helps fill out the gaps in the kick drum's sound.

- If your bass sounds too thin then adding some 200-250 Hz can beef it up.

- Another good way of fleshing out the bass is to add smaller frequency boosts in multiple frequency areas instead of one large boost in the lows. If the fundamental frequency of the bass (or the key of the song) is 100 Hz then adding a boost at 200 Hz, 400 Hz, and 600 Hz (various multiples of the original frequency) will flesh out the sound of the bass more than just one large boost at 100 Hz. Harmonics can be your friend and can add extra depth to your EQ'ing.

- In a similar vein, getting the bass to cut through isn't only achieved by boosting the thickness in the low-mids. Adding a boost at 600 - 800 Hz can bring out the upper range of the bass, giving it more presence in the mix.

- The bass can also get in the way when you least expect it. Sometimes a muddy vocal is just the bass masking the vocal track, so make sure your bass isn't cluttering up some part of the mix that you're not thinking of. A good way to check this is to boost the bass in the mids and upper-mids and listen to the "other" tracks, the vocal for instance, to see where the bass starts masking those instruments.

■ GUITAR

- Don't assume that the killer guitar tone you got during the recording will fit with the rest of the arrangement.

Sometimes you'll need some serious sculpting and cutting in frequencies you wouldn't expect in order to make the guitars fit both with the other instruments as well as fit within the genre (more on this later).

- Make sure your guitar isn't clashing with the bass in the low-mids. Cut out a little bit in the 150 - 250 Hz region if either instrument lacks definition down there.

- If your guitar is sounding thin and it's not getting in the way of the bass, a boost in the low-mids around 200 - 250 Hz can help bring some thickness to it.

- If the guitars are lacking body and power, increasing the area around 500 Hz can make them bigger without making them sound muddy or hissy.

- If your guitars are overly distorted and hissy, reducing the high-mids around 4 kHz can clean that up.

- The high-mids from 1 - 4 kHz are a very interesting area to experiment with, especially when it comes to rock guitars. You can really shape the tone of the guitar to fit the style of playing and the genre of the song just by picking the right high-mids to boost and cut. Spend some time getting familiar with this area and you'll be a guitar EQ virtuoso in no time.

- If you want to separate two rhythm guitars that are doing the same riff then try boosting a flattering high-mid frequency in one guitar track and cutting it in the other. Then find a separate flattering high-mid frequency in the track you just cut and repeat the process.

- If you're EQ'ing an acoustic guitar and it's sounding a little too honky and "cheap-sounding" then hunt around and cut in the 800 Hz region to give your acoustic a smoother sound.

- You can bring out some brilliance in an acoustic guitar by boosting around 8 kHz.

- However, if your acoustic guitar is just a small part of the arrangement and just needs some "strummy presence" in the background then adding some air above 10 kHz with a heavily high-pass filtered track can help the acoustic feel present in the mix without cluttering up the rest of the arrangement in the lower frequency spectrum.

- If your guitars sound like they're masking the vocals a wide cut around 1 - 5 kHz can help settle the guitar around the vocal sound, letting both tracks fit together in the mix.

- Even if you have wide cuts, like in the previous tip, you might want to accent a specific frequency inside that cut to bring the guitar (or any other instrument) out in the mix. You'll end up with something that looks like a wide cut (a valley if you will) but then you have a boost inside it where you accent that particular frequency. Then your EQ curve will look like a wide valley with a little mountain inside of it.

■ KEYS, PIANOS AND SYNTHESIZERS

- Many of the tips I've talked about above can be used on keys and pianos. Fullness around 200 Hz, presence at 3 - 5 kHz, and air from 10 kHz and above.

- Keys can quickly clutter up the mix so be mindful of the low-mid frequencies. Add a shelving cut to tame the lows if they get in the way of the kick, bass, or guitars.

- The main tip when it comes to keys is to listen in the frequencies you're not targeting with the other instruments. I've found that the mids from 600 Hz - 1 kHz are often underrepresented in the rest of the arrangement so I tend to focus on that area to bring out the keys and pianos in my mixes.

- Other frequency ranges like 300 Hz can often be helpful, even if they're mostly associated with "boxiness." Don't get too caught up in what each frequency range is "supposed" to sound like because the only thing that matters is whether your EQ adjustments are making your instruments fit together and cut through the mix.

- If your piano is getting in the way of the vocal, do the same EQ cut trick in the high-mids as you would with the guitar.

- You'll want a piano to sound lush and big when it's playing on its own, but make sure you EQ it in the context of the mix if there's a big arrangement going on. It might not sound good in solo, but the only thing that matters is how it sounds in the overall mix.

- Because of the near-limitless sonic potential with synthesizers, it's hard to give any concrete guidelines. You'll find muddiness or boom in the lows from 100 - 200 Hz, you can add thickness in the low-mids to middle frequencies from 300 - 600 Hz, and you can adjust attack and presence around the 1.5 and 3 kHz areas respectively.

■ VOCALS

- If the vocal is lacking clarity while still having plenty of high-end energy then reduce the 200 Hz area to clean things up.

- If your vocal sounds too nasally then look at the 900 Hz

- 1 kHz area to cut it out.

- The area above 1 kHz, around 1.2 - 1.5 kHz can often help the vocal cut through the mix. Just make sure you don't accidentally make the singer sound nasally!

- I've often found that when I'm trying to reduce honkiness in the 1.2 kHz area I sometimes fail because it's often better to reduce the 300 Hz "boxy" area instead. That can fix the entire vocal sound.

- Bring out the clarity and intelligibility of the vocal in the 3 kHz range.

- A 5 kHz boost tends to add a nice presence to the vocal, but boosting too much of any high-mid frequency can quickly result in overall harshness.

- Sibilance (the harsh 's' sounds) is generally centered around 7 kHz. However, certain 's' sounds can be sibilant at lower frequencies, and if you have particularly problematic sibilance, you might need to hunt for their harmonics too, sometimes all the way up to 14 kHz.

EQ is a big part of mixing and almost too big a subject to condense down to one chapter in a book. EQ is subject to taste, experimentation, and style that changes with every mix you do. Every time I open up a new mix I instinctively think of these guidelines when I'm listening to what I want to add or subtract.

However, that doesn't mean I blindly follow these suggestions if those decisions don't make a good mix. I've often had to fly in the face of common wisdom just to make things cut through and fit together, and that's always what's most important when it comes to mixing. Nobody cares that you can pinpoint frequencies like a wizard. People care whether the mix sounds good, nothing else.

A NOTE ON REBALANCING

As you move through the mixing process and keep adding processors such as EQ and compression, you will inevitably change the initial balance you made with only volume and panning. That's why it's always a good idea to keep rebalancing the faders as you move through the mix. Even if you keep the gain structure of the plug-ins relatively perfect (equal level going into the plug-in as is going out), you will still need to keep adjusting the volume. Mixing isn't a complete step-by-step process. It's more a set of steps and guidelines you follow while constantly adjusting and reacting to your mix decisions.

Before we end the chapter, I want to recap what we learned about EQ:

- EQ is your best friend when you're trying to separate your instruments in the frequency spectrum.

- However, some issues can't be fixed with EQ because of their dynamic nature. A track that's constantly switching from quiet to loud can't be tamed with EQ. That's where compression comes in.

- You don't _always_ have to use EQ on every track. Some tracks might sound great as they are and only need some compression and effects to fit with the rest of the mix. Maybe all they need is a little filtering just for low-end control. So remember that you might have tracks in your mixes that you don't feel the need to EQ. _That's ok!_

- The mixing process goes much faster if you adopt bus processing on groups. You might not be comfortable with the top-down or middle-out method right away, but it's worth it if you want to save time and do more mixing.

- After EQ adjustments (and other mixing techniques) you

will often need to rebalance the faders so that the instruments all end up where you want the balance to be.

It's my hope that the tips above have given you some good ideas on what to try in order to create separation between your instruments and definition in your mixes.

To be fair, the art of EQ is an important topic, and it's something you should become intimately familiar with. A short chapter in this book is probably not going to make you an EQ pro. In fact, I wrote an entire guide with in-depth EQ tips and strategies. You can find links to more EQ resources at www.StepByStepMixing.com/Resources page or go to www.EQStrategies.net to check out *EQ Strategies – Your Ultimate Guide to EQ.*

Chapter 4 –
Using Compression

Alright, now let's get started on compression.

Compression is the second-most important way to process audio after EQ.

However, compression is a bit more complicated than EQ, it's also more subjective to the person using it.

If your mix is muddy, it's just muddy in general to everybody most of the time. There are a few ways to fix that, but it revolves around cleaning up the low-mid area. That's a pretty simple job for your EQ.

But compression is different.

Compression is highly subjective, and you can use it so many different ways. You can use compression five different ways that all sound different but still GOOD. It just depends on the style you're going for, and the sound you want out of your mix.

WHAT DOES A COMPRESSOR DO?

You've probably heard that professional mixing engineers use compression when they mix. Understanding what a compressor does and when you should use it might give you a clearer insight into why you should compress your tracks and how to go about it.

In the simplest of terms, a compressor is an automatic level controller. It automatically makes the softer signals louder and the louder signals softer. As such, it manages your signal level for you, raising it and lowering it depending on how loud it is. But using compression and knowing how to compress are two different things.

■ WHY COMPRESS?

Instead of slapping a compressor plug-in on each track because that's how you think it's done, ask yourself what the end goal of using the compressor is going to be.

- Do you need to tighten the drums?
- Do you need to level the vocals?
- Do you need to tame the attack of the transients?

All of these questions have different compression answers depending on what you're trying to accomplish. It comes down to the mentality of using compression as a tool, not as a crutch.

A good rule of thumb when using compression is to ask yourself:

"Why am I compressing this track?"

Always have a reason for compressing something. If you have a problem with the peaks then know what you're targeting when you add a compressor to the track, like with the vocal tip above. If you need to add thickness to your drums, know how you can go about it with high ratios. If you throw compression on a track and it doesn't sound better no matter what you do, then maybe you should just leave it off.

HOW TO APPROACH COMPRESSION

Let me tell you how to approach compression by telling you a little story about gambling at a Vegas Blackjack table.

I'm not a high-roller, but I do like gambling, especially at games like Blackjack where you have a system and don't just rely on blind luck. Blackjack is a great game to play if you know how to play by the book. You can play for hours if you just follow the simplest of systems.

For instance, if the dealer has a 5 and I have a 12, then I stay and don't ask for another card. There's a higher probability of him busting, so I'll stand my ground and hope the probability plays out. You always hope the dealer busts, and when he has a low card, there's a higher chance of him losing.

I also always hit on 16 whenever the dealer has a 7 or higher. You always assume that the dealer's other card is a 10, so by default you've lost. But if you hit and get anywhere from an Ace to a 5, you're better off. I'd rather hit and lose than wait and lose any-way. And finally, always split into Aces and 8s and double down on 10s and 11s if you're higher than the dealer.

Why did I tell you all this?

It's a system. A system is a set of guidelines and rules that I've memorized. It's actually a pretty great system that works most of the time. You don't always win, but you can end up playing for a long time without losing any money.

USE A COMPRESSION SYSTEM

Using compression works the same way. Just like in Blackjack, the system I use with compression has a greater probability of working, so I stick to it and tweak as needed.

I follow my go-to presets in my head whenever I'm mixing. A kick drum starts with a ratio of 4:1, a gain reduction of 3-6 dB, and a medium attack and release. With just a few tweaks from there, I'm able to get the sound I want.

Bass guitars start with the ratio relatively high, the attack fairly fast, and more gain reduction than other instruments just to get it fat and steady. Sometimes that works well, but other times I need to move things around to get the bass to sound right.

And finally, I always parallel compress the drums and shape the snare compressor in time with the track. Just like doubling down on 11 is the smartest move when you find yourself playing Blackjack, so is buss compressing on the drums.

It just works 90% of the time (it actually doesn't work 90% of the time to double down on 11s, but you catch my drift). However, for drums, if you want that punchy drum sound without overpowering the rest of the mix, that's the way to go. There's a system for everything, including compression. My system helps me work faster since I know from experience what usually works.

It's perfectly fine (and fun!) to experiment with different compressor settings when you're learning to use them, especially when you're figuring out which compression style works the best on which tracks. But in the end, having a plan and a goal whenever you're adding compression to a track is more efficient and effective when you're working for clients and not just for fun.

So when your tracks are nicely balanced and EQ'd, the thing you reach for next is the compressor. Only this time you should ask yourself why you're doing it and then try to accomplish those specific goals.

Once you know that you can start playing with your toys because now you have a goal in mind.

COMPRESSOR PARAMETERS

First things first, let's talk about the common parameters and knobs you'll find on your compressor.

Most compressors will have a variation of the following parameters. However, some will be incredibly simple, like any compressor that emulates the LA2A with its two-knob compression system. If you're working with simple compressors such as that one, I really only have one rule for you to follow:

"Tweak it until it sounds good!"

It's really as simple as that, but if you're working with something a little more complicated here's what you need to know:

- **Threshold** – The threshold basically sets the level for where the compressor should start listening to the audio signal. If the audio is very low in level and the threshold is too high it won't "hear" the signal so it will effectively be useless. Any other parameter doesn't matter because the compressor doesn't start working until the audio hits the threshold. Make sure the audio signal actually reaches the threshold so that it actually starts compressing.

- **Ratio** – This is the "amount of compression" that takes place once the audio signal crosses the threshold. So if the input level goes over the threshold and the ratio is

2:1 then it will divide the level above the threshold in half and compress it down by two. So the higher the ratio, the more extreme the compression. 10:1 and higher is usually called limiting because any signal over the threshold gets compressed so hard that it gets pushed down to where the threshold sits instead of letting some of it through.

These two parameters work in tandem. Many compressors only have these parameters and nothing else.

You want to know how everything works so let's check out the next set of knobs here, the attack and release.

- **Attack** – This is basically the amount of time you give the compressor before it reacts to the incoming signal. If the attack is fast it'll compress immediately after the input crosses the threshold. If the attack is slower it will take a while to react. Picking the right attack time is crucial for shaping sounds because it can change the way the transients of the audio signal are affected.

- **Release** – The release time determines when the compressor stops compressing. A faster release time means a faster recovery time for the compressor. A longer release time means that it keeps the signal compressed for longer, which can result in pumping with very rhythmic signals because the audio is never allowed to go back to its uncompressed state.

These four parameters are the most important ones to keep in mind on any compressor.

Other compression parameters include:

- **Knee** – Selecting a soft or hard knee will change the way the compression is applied, either gradually as the signal approaches the threshold, or linearly as soon as the audio hits the threshold.

■ COMPRESSION STYLES

There are a lot of different compression styles, all of which work for different instruments or for specific compression characteristics.

Let's talk about some of the most common styles:

- FET
- Opto
- VCA

If your compressor has a FET (Field Effect Transistor) mode or style it's basically emulating an 1176 compressor. The 1176 is perhaps the most famous FET compressor. People like to use them to get punchy drums.

An "opto" style emulates the LA-2A optical compressor. It doesn't react as quickly to your audio. It works well for parallel compression if you don't want an aggressive attack/release, which we'll discuss in depth later on in this chapter.

The VCA style is fast and transparent. The VCA style doesn't color the sound as much as the other ones, so they're ideal when you want your compression to go unnoticed. One of the most popular VCA compressors of all time is the DBX 160.

Compressors tend to build upon these models. There are other emulations, but these are the most common and popular out there. They have a specific sound or character that's different from generic, stock models.

It's no surprise that an LA-2A sounds different to an 1176 in your mix. The "circuitry" is completely different. That's why many all-in-one compressors have a "type" button that allows you to change the character of the compressor.

Once you've decided what goal you're trying to achieve with your compression, the next question should be what type of compression do I want to use. Depending on the style of music and instrument you're running through the compressor, these styles will change the way the compressor works on your audio, so it's worth experimenting a bit until you find the style you like the most.

- **Input/Output** – This is important for gain staging purposes and obviously changes according to the level of the signal you're feeding into the compressor. If you have a very quiet signal you might need to put the gain up a bit so that the compressor works better (or down if you can't put the threshold low enough and it's always compressing way too much). A good rule of thumb is to increase the output according to the same amount you're compressing, but I like leaving it on auto. Also, whenever you're compressing make sure that you don't get drastic level differences when you bypass the compressor because it'll make you think everything sounds better when it's actually just sounding louder. If you can see the input and the output of the signal side by side it's a good idea to trim the output so it matches the input. If you trim the input you'll screw with your threshold setting because the signal coming into the compressor won't be as loud. Honestly, when in doubt, auto-gain is a useful feature that helps you set it and forget it!

- **Dry Mix** – If you have a dry mix knob it allows you to add in a bit of the uncompressed signal, which can come in handy if you want to do direct parallel compression without using sends and busses.

- **Metering Window** – Usually you'll have some sort of metering window. It's handy for seeing the waveform on the screen, what the compressor is doing to the signal and for seeing how much gain reduction is going on. You can often switch between input, output and gain reduction to see how your compressor is affecting the signal.

Typical compressor parameters. Logic Pro stock compressor.

APPLYING COMPRESSION

Compression is a never-ending subject for me. I always feel like I learn new ways to use compression with every mix I do. Sometimes it's as simple as slapping on an LA2A and calling it a day, sometimes I wrangle with the attack and release for way too long without getting the punch I want out of the instrument.

The following practical tips are therefore just a series of things that have worked for me, as well as a few things I tend to avoid.

■ PRESET MIXING

A technique that can often help to speed up your mixing is starting with presets. I often pick a preset that seems to offer what I'm looking for and then I tweak it according to the song and the track. Obviously the preset designers haven't listened to the song I'm mixing, but starting with a preset puts me in the ballpark of what I'm going for. You'll usually have to tweak the threshold at least and then adjust the other parameters to taste.

■ SUBMIXING

If you've followed my method of top-down or middle-out mixing in the EQ chapter then you might be wondering where to put the compressors. What I tend to do is put compressors on each submix and group to glue all the tracks together. Then, if the individual tracks need some extra punch I'll experiment with more compressors there. The trick to using a lot of compressors throughout the signal path is to not compress a lot at each stage. So aim for a few dBs of gain reduction on the groups to tighten them up, and then use your own judgment on the individual tracks as to how much more gain reduction you need to apply.

■ DRUMS

Starting with the drums I like adding some punch to the kick drum. Higher ratios, starting at about 4:1, give the kick more thickness because it pushes the entire level down and thickens it up.

However, if the compressor starts adding too much thickness in the low-mids you might need to compensate by cutting some of that out with EQ.

When you're compressing a kick drum pay special attention to what the difference in attack and release does to the transient of the kick drum. For instance, a medium attack and release time lets the initial beater of the kick through without the compressor dulling the initial transient. With a medium release timed more or less in time with the song you should be able to hear (and see on the meters) the compressor reset itself between hits.

The same goes with the snare. I tend to choose a few different compressors, either a few different plug-ins or just a few of the emulation styles inside the stock compressor in Logic. Each has a different sound and depending on the genre, the sound of the snare and what I'm going for, one of the styles usually complements the snare better than others. Just like with the kick drum, be aware of the attack and the release times because you can really dull the thwack of the snare if you're not careful.

One of the things you need to think about when compressing any individual drum track is the bleed from other drums. For instance, if you compress the snare a lot you risk raising the overall level of the snare. What that means is that there will be more kick and hi-hat sound in the snare mic if you recorded a live drum kit. We won't be covering gates or downward expansion in this book but it's a great way to reduce the bleed between instruments and can really help cut out a lot of noise in your tracks.

On the drum bus itself I usually go for a smooth feel, just making everything a little tighter and glued together. I'm careful not to make the attack and release times too dependent on one drum, even though it's obviously triggered mostly by the kick and snare. Beware of making the release of the compressor too slow because you'll introduce a pumping effect to the signal which is when it never gets to reset itself between hits.

Lately, I've found success on the overheads by using an opto-style compressor, such as an LA-2A, and just letting it ride the gain reduction a few dBs. It seems to thicken up the overheads and usually makes the cymbals cut through the mix without being too abrasive and forward.

The fairly fast release lets the cymbals have their place without pushing them way down, which would sound unnatural.

If you have a room mic on your drum kit you can usually have a bit of fun with it. Use it as the "crush" mic, where you can get pretty heavy-handed on the compression. If you really crush this mic with a high ratio, lots of gain reduction, a fast attack and long release, you'll have a terrible sounding track on its own but it can sound great blended in with the other drum tracks to give some additional depth to the drum mix.

BASS

I like keeping the bass steady with compression. That means a high ratio and constant gain reduction. Of course, this is always dependent on the genre, but if the bass has more of a supportive character rather than a lead role, then keeping it tight in the background is what I usually go for. It's often important for the bass to keep the arrangement steady so I use the bass to anchor the arrangement. Because of this, I don't want drastic dynamic changes throughout so I regularly use a high ratio to keep it thick and steady. However, if the bass is played very steadily it might not need as much compression, although a high ratio can still keep it grounded. You might think a high ratio would always make things sound squashed, but it's really a combination of multiple things, especially the attack and release times. With an attack setting that lets the initial transient through you can get away with compressing the signal more without it sounding squashed.

Sometimes, for bass and other instruments you just need a gain reduction of a few dBs to keep everything steady. It just gives the whole bass a little more thickness. And running through each of the styles, I settle on the style that seems to fit the track the best by, you guessed it, using my ears and listening to it in the context of the mix!

■ GUITARS

If you're dealing with strummy acoustic guitars you might want to control the peaks while leaving the rest of the signal mostly alone. If that's the case I recommend picking a medium ratio at 4:1. Then, try to find the threshold spot where the gain reduces about 2-3 dBs at all times while pushing the peaks a little harder in those spots where the accents come through a bit more. When you're starting out learning about compression it's good to have a visual display in your compressor so you can see what's happening to the waveform. Some stock compressors have this but one of my favorites is the Fabfilter Pro-C because you can really see what parts of the waveform the compressor is affecting. That way you can zero in on the peaks while leaving the rest of the signal fairly intact.

In order to catch the peaks you'll need a fast attack. If your attack settings are continuously variable (and not just Fast/Slow) then you have some control over how much of the transients you let through. If you like the strumminess but you want to catch most of the peaks, then gradually moving the attack faster and faster while listening to the signal will tell you where the most optimal millisecond setting of the attack time is. The thing about strummy guitars is that they cut through the mix. If you set the attack time too fast you might not end up with something that sounds good in context. For that reason you'll need to really listen to what you're doing as you're shaping the signal with compression. Analyze the instrument in solo, but mix it in context with the rest of the arrangement.

For electric guitars I generally tend to cheat and use a plug-in like the Renaissance Axx that only lets me control the attack and the threshold. Plug-ins that are specifically designed for guitars in mind are a great way of taking out the guesswork. And if they don't sound good, I can always move onto something else. Low ratios of around 2 or 3:1 can help tame the guitars and generally tend to thicken things up without squashing the signal. Combined with a few tweaks of the attack and release times and aiming for a gain reduction of a couple dBs, you usually end up with a nicely compressed guitar track that sits well in the mix.

■ VOCALS

If a vocal has a lot of dynamic range there's one thing you can do, similar to what I recommended with the acoustic guitars. Keep the threshold very low so it's hardly compressing and use a high ratio so that the compressor only reacts when those loud peaks go over the threshold but leaves the rest of the signal intact. A fast attack and release time should keep the compressor working only on those peaks. A FET-style compressor would work well in this instance because it's a pretty fast style and has a hard knee so it squashes that initial audio. You'll be able to see the compression working only on those peaks, which controls the vocal a bit better in those parts.

An alternative way to compress vocals is to do the exact opposite. Use a low ratio of 1.5:1 or 2:1 and set the threshold so it's continually compressing. This squeezes the vocal and keeps it tighter without making it sound over-compressed. You can do both in series, which is similar to my next trick.

■ COMPRESSION IN SERIES

You don't have to use only one compressor or compression setting on a track. Putting two compressors on a track in series can actually help if the compressors are tackling two separate problems.

If you want to have the vocal sit better in the mix after dealing with the hypothetical dynamics problem above you can add a second, slower compressor after it. An LA2A style opto compressor with a soft knee, a 2-3:1 ratio, fast attack, and medium-ish release will help the vocal sit better in the mix. Then it's simply a matter of bypassing the compressor to hear if the vocal sounds better. Usually you can tell because the vocal sounds weaker and isn't as present and level in the mix.

This series compression tip is very similar to the tactic of using the 1176 compressor to tackle the peaks and the LA2A to take care of riding the vocal. It's actually very common and you can get the same effect by using the FET style and Opto style in any of your compressors (assuming they have multiple styles).

■ BACKUP VOCALS

Backup vocals can be tricky because you want to keep them audible in the mix but out of the way of the lead vocal. One way of doing that is with EQ by cutting the high-mids from 900 Hz - 3 kHz to create a pocket for the lead vocal, but compression also plays a part. I tend to group all my backing vocals and then treat them with one bus compressor. If you find a nice middle ground where all the vocals stay consistent and level you should be good to go. Usually there's a good preset to use as a starting point, and then you just tweak it according to what the backup vocals are actually doing. A ratio of up to 4:1 is fine because you usually want to keep things a little tight in the background. Combining that with a fast attack means we don't get any pesky transients escaping through to clutter up the rest of the mix. Tweak the settings, solo for analysis and then play it in context with the mix. If the backup vocals jive well with the lead vocal, you're good to go.

■ PIANO AND KEYS

Compressing keyboard and pianos is something I've never really had a method for. The difference between a gritty electric keyboard and a spacious grand piano means that they require two drastically different approaches.

For rock keyboard and organs I like pushing them a little hard to get them to cut through the mix. Especially if it's not really acting as a lead instrument and can stay in the background, a higher ratio can keep it underneath everything. Then it's a matter of tweaking the attack to let the initial transient cut through because a high ratio can tend to squash the signal pretty heavily. Then you just tweak the threshold until it sounds like it's compressing nicely and sitting well with the backing track.

When it comes to pianos you don't want to go overboard in compressing them because you want to retain the natural dynamics of the instrument. That said, you do want to tame the peaks, especially if the piano is playing aggressive chord stabs (think "Love Song" by Sara Bareilles). At that point a fast attack and a medium release will work well. A 3:1 or 4:1 ratio is a good middle point between taming the peaks and squeezing the dynamic range.

PARALLEL COMPRESSION

A great way to add a bunch of punch to your mixes is to use heavy-handed parallel compression alongside the compression you add to the individual tracks themselves. You may only want to gently ride the gain reduction on the tracks, but if you add parallel compression on a separate track you can get very creative.

How you do it is simple. In this case let's imagine you're adding parallel compression to drums to get them really rocking.

1. Send your drum bus to another aux track.

2. Add a compressor to the aux track.

3. Hit the compressor hard, with a high ratio, lots of gain reduction, a fast attack, and a slow release. You want the drums to be really compressed and pumpy on the aux track.

4. For added measure, add an EQ to the aux track after the compressor. Boost the highs and the lows. This makes the highs and lows more present without adding to the

transients of the drum track. Because the attack is so fast and the release is so slow the compressor eats all the transients up so that all you get is this thickness in the lows and sheen in the highs.

5. Blend the compressed aux track with the drum bus to taste.

Usually, this results in a sound that cuts through the mix without getting in the way of the rest of the instruments. It just adds extra presence to everything.

You don't need to limit this technique to only drums. Try it on vocals to get them very present in the mix. I've had great luck with multiple types of parallel processing on various instruments, so if a track is lacking some punch but it reacts poorly to your compressor on the insert then try it in parallel to see whether it adds the punch you need.

WHAT ABOUT MULTIBAND COMPRESSION?

A multiband compressor splits your audio into different frequency bands, while the rest of the compression parameters stay the same. It's great for when you want to compress a particular frequency range differently to another.

I was afraid of using it for the longest time because I wasn't sure how to work it correctly. So I just kept on using my other compressors, not realizing the potential I was passing over.

I always used to use them in mastering, but that somehow seemed easier and more straightforward.

It wasn't until I realized the problem-solving capabilities of the multi-band compressor that I finally got hooked.

Here are just a few of the ways you can use a multi-band compressor:

- **On the master bus** – Getting a great mix from the top-down with multi-band compression lets me dial in the right compression in the right frequency ranges.

- **To tame muddy bass guitars** – I sometimes slap a multi-band compressor on a bass and compress a little harder on the low-mids to get the bass to sit better in the mix without adding too much boominess.

- **On the drum bus** – A multi-band compressor lets you compress the entire drum-kit at varying levels throughout the entire frequency spectrum. This is great when you need to hit the kick pretty hard in the low-end but don't want the cymbals to constantly compress every time the kick hits. Conversely, you can also tame harsh cymbals this way if your cymbals are too present, in addition to controlling pesky resonances and ringing overtones from the drums.

- **Smoothing out vocals** – If you find a specific frequency in the vocal that's causing problems just tweak the crossover points of the multi-band compressor so that it's taming that particular frequency range while leaving the rest of the vocal alone.

The multi-band compressor is a seriously useful tool to make better mixes. These are only a few of the ways you can use it to tighten up your mixes so make sure you experiment with other instruments and tweak the frequency bands accordingly to taste.

Given a good starting point, a great result is just a few tweaks away!

REVIEW YOUR MIX AND REBALANCE IF NEEDED

Your entire approach with compression basically comes down to this:

1. Add compression to any track that needs it (probably almost all of them).

2. Find the style of compression you like.

3. Find a preset if you want a simple starting point.

4. Adjust the threshold and the ratio to suit your needs.

5. Tweak the attack and release times to shape the dynamics of the signal.

6. Analyze in solo if needed, but always listen to the track in the context of the mix.

Once you've added compression to every track that needs it and you've made your mix punchier and more manageable it's time to rebalance. Chances are you've overlooked some proper gain-staging (it happens to everyone) so you might want to go back to your faders and rebalance any instruments that have become too loud or too quiet. Additionally, compression can negatively affect any EQ boosts you made before adding compression so you might want to add an EQ after your compressor to compensate for any frequency loss.

After you've balanced, EQ'd, and compressed your mix you might notice that the arrangement all sounds a bit flat. That's not surprising at all. You haven't added any depth to the mix yet. In the next chapter we'll talk about how to flesh out your mix in the space and depth department using reverb and delay.

So to recap, the main things to think about when it comes to compressing are:

- Make sure the threshold is relative to the actual signal. A threshold that's set at a point where the audio never crosses the threshold won't make a compressor do anything.

- Tweak the ratio depending on how thick you want your sound. I've found that lower ratios work better on things you don't want to have sound too compressed but higher ratios are good on things such as kicks, bass, and other things you want to sound punchy but don't care about reducing the dynamic range too much.

- The attack and release time settings can really shape the sound of the instrument. Slower attacks are good for when you need the transients to cut through. Slower releases can tame the sustain of a signal.

- Lastly, have a reason to compress, even if that reason is "I wonder what an LA-2A sounds like here." If you're just slapping a compressor on everything, not tweaking any settings and thinking that a compressor on a track is required then you're not using compression to its full potential. Tweaking the settings, modifying the threshold, ratio, attack, and release can really help you shape your mix, so spend some time with it. Even cranking up your instrument in solo just to hear the minor details that the attack and release contribute can be very educational.

Chapter 5 –
Reverb and Delay

Now we're going to talk about adding space to your mix.

Hypothetically, if I was mixing along with this book the song would be sounding fairly balanced by now. You would be able to hear everything that was going on, but it might sound a little flat. That's just because most of the instruments are bone dry and all occupy the same region, sounding very two-dimensional.

The mix is panned so there's stuff from left to right, and it's EQ'd and compressed so all the instruments sort of work where they are now, but there's no front-to-back spacing. It's all in the front so it's hard to enjoy listening to it like this.

The concept of space in a mix is very subjective, so this will be an interesting chapter. You might not use every single tip in this chapter, but pick and choose according to your needs. Otherwise you might derail the mix with too much reverb. You don't want your mix to sound washed out so make sure you're careful that you don't get too heavy handed with adding space.

There are so many different ways to use reverb and delay. I'll try to give you several different options that you can use so you'll end up with something interesting at the end. Just make sure that you don't get too carried away with the experimentation. And reel that reverb in at the end so it doesn't sound too amateur!

REVERB PARAMETERS

In every reverb engine we have some of the same parameters to fiddle with. The more basic reverb plug-ins only have one or two, but some of the more advanced ones have almost unlimited capabilities for customization. Let's look at some of the typical parameters we usually see in a reverb.

- **Room Size/Type** – The size determines how big your reverb will feel and how big your mix will sound. This can either be predetermined by the reverb type/mode or just the number of seconds you make the decay of the reverb. For example: a 0.5 second reverb sounds much shorter and smaller than a longer and lush 3.3 second one.

- **Pre-delay** – A pre-delay pushes the reverb from the source sound. It essentially delays the sound of the reverb by the number of milliseconds you choose. Think of pre-delay like the distance to the walls. With more pre-delay the walls are farther away, which means there'll be more time before you hear the reflection coming back to the sound source. It's great for making a big ballad sound spacious while still keeping the vocal up front. Tweak pre-delay by 20-40 ms to hear how the reverb pushes back away from the initial phrases of the vocal for instance.

- **Early/Late Reflections** – These are the sum of all the reflections that happen after the source signal. If you think about yourself in a room, the early reflection is the first sound that bounces back from the nearest wall. The early reflections can easily tell us how big or small a room is. If the reflection happens almost instantaneously then we're in a small room. If it takes a while to get back to you, chances are you're in a cathedral or a large hall.

- **Damping** – This parameter dampens, or softens, the higher frequencies. You don't always want a really bright reverb because it can be very distracting in the mix. Think about damping as the natural stuff you have around you, the curtains, carpets and the couches that absorb the higher frequencies. A damping factor on a reverb basically determines how much of those highs are filtered out. It works fairly similar to an EQ. High frequency damping cuts the highs, low frequency damping eats the lows.

- **Density** – Density decides how "thick" your reverb becomes. The denser it becomes the more the reflections continue to pack together to create a thicker sound. If your reverb has a lower density, you'll create more space and time between each of the reflections. At the low extremes the reverb will sound more like echoes than a natural reverb that blends with itself.

- **Diffusion** – A diffuser basically scatters the reflections and makes the room sound more live instead of harsh and reflective.

- **Frequency Filters** – This is fairly self-explanatory. EQ'ing your reverb is incredibly important to shape it into your mix. Sometimes all you need to do is filter the highs and the lows to get most of the space in the mids. If your reverb doesn't have frequency filters just add an EQ after the reverb plug-in on your aux bus and use that to control the frequency response of the reverb. Too much low-end can clutter up your mix in no time. Conversely, high frequency reverb can sound hissy, sibilant, and harsh, none of which are good adjectives for your mix.

- **Wet/Dry Mix** – It's a good idea to use reverb and delay as send effects rather than inserts. That way you'll be able to treat them separately and blend the instruments better together. If you don't want the hassle of using sends then you can insert a reverb on the track itself. As an insert, put the dry mix to 100% while tweaking the wet mix until it sounds nice and "reverb-y." Otherwise, if you use the reverb as a send make sure the reverb is set to 100% wet.

THE DIFFERENT REVERB MODES

All reverb modes sound different. The spacious Spanish Cathedral is going to sound a lot different than your living room.

The bigger the spaces are, the bigger they sound. We engineers know what we can accomplish musically by harnessing these spaces in our mixes. The different spaces we can choose from in our productions are called "room modes". Room or reverb modes are basically categories of different spaces that have a distinct character and sound.

- **Room** – Small rooms, low ceilings. Think a garage, bathroom, or small studio room. It's nice to add a little ambience to instruments. It's also nice if you find a good room sound that works to glue an instrument section together. It's never going to sound as lush as a hall, but it has the right character for some genres and sounds.
- **Hall** – Halls are big and lush and sound great because they're usually designed after fancy concert halls. However, small halls can feature the best of both worlds:

lushness while staying tight. Use big halls to create big vocal productions, huge drum sounds, and other things you want to make big in the mix. Small halls still sound bigger than small rooms, and they have a more reflective quality than a room.

- **Chamber** – Back in the old days the engineers would send the audio from the control room to be played back on speakers inside large reverberant chambers. They were essentially the first real, constructed reverbs. Microphones would pick up the ambience and reverb from these chambers which would then be added to the mix. Chambers can sound big but usually do not have many early reflections, giving the sound more space without audible reflections.

- **Plate** – Plates are very interesting contraptions. In the real world, they are big electro-mechanical plates that vibrate with the music. The pickups on the plates pick up the reverb vibrations that are then added to the music. Plates are usually pretty dense but do not sound roomy or echo-y, with a smoother reverb sound than some of the other modes. Plates are popular for drums (especially snares) as well as vocals. Of course, these days we just work with the software emulations of these big mechanisms. One of my favorite plug-ins is the Universal Audio EMT 140 Plate Reverb.

- **Spring** – Crank up the spring reverb on your guitar amp and kick it! You'll hear a big "Boing!" That's spring reverb. It is very popular in surf music and great for guitars. Sound is sent through springs which reverberate and create that boingy, springy sound. Although mostly reserved for guitars, don't be afraid to experiment with it on other elements in your mix.

- **Impulse Responses** – Impulse Responses (IRs) are digital snapshots of acoustic spaces. Some reverb plug-ins are IR-only (like Logic's Space Designer). They usually have plenty of cool sounding impulse responses ranging from simple rooms to weird spaces commonly reserved for space design (think ambient noise in a space movie). Although creating impulse responses is outside the scope of this book you can find a lot of custom IRs with a simple Google search.

AN EASY WAY TO FIND THE RIGHT REVERB FOR YOUR MIX

With all the options of reverb types, it can be hard to know what you should choose for your next track.

What should you think about when choosing the right space for your song? Even though you have a favorite sounding reverb it might not work for that alternative Folktronica act that you just recorded.

- **Think about the tempo of the song**

  If you are working with a fast tempo song that needs to come across clearly and well defined, mucking it up with long reverb will just make the mix sound cluttered. Think about the tempo of the song and select your reverb accordingly. A slower song can use longer reverb, and faster songs might need really short reverb. Or you could just use delays instead.

- **Think about the wetness of the song**

  Do you want your overall mix to be dry, or do you want lush reverb filling up the space? Are some of the pre-recorded tracks already rich with heavy reverb and space, or was it all recorded extremely dry? Think about how wet you want your mix to be and choose your reverb according to that.

- **Think about the lushness of the arrangement**

  Is the arrangement going to be really dry and in-your-face, or is everything going to be drowned in space? You might need to keep some elements dry even though you are going for an extremely wet mix. The more reverb you are putting on your instruments the more you need to anchor it down with at least a few dry instruments. Also, be wary of adding too much reverb since it can be a sign of a very amateur production.

- **Think about the rhythm of the vocal track**

  Is the singer singing long sustained notes, scat singing, or rapping? These are some of the factors you might want to consider when choosing your vocal reverb. The rhythm of the vocal can tell you if the reverb you've chosen actually works. If you're going for "My Heart Will Go On" Celine Dion long sustained singing then a large and long reverb might work exceptionally well. But if your singer is Scatman John then a long reverb will probably just get in the way. Try a short plate mode, ambience reverb or delay for more staccato styles of singing.

- **Think about the room**

  Think about the instrumentation and experiment with placing them all together in the same room. Some instruments sound great together when they're "placed" in the same room or reverb plug-in. Say you have a great sounding impulse response from an old recording studio. Use the reverb of that room to glue the instruments together.

- **Think about contrast**

  Use different reverbs for different parts of the song to create contrast.

  Let's take the drum reverb for example. The verses might be big and spacious, with heavy reverb, floaty, and crisp with the cymbals. Then create a punchy chorus with parallel compression and a small drum room to create a tight contrast to the floaty verse.

  Take a moment to think about the characteristics of the mix at hand and then choose the reverb you think is right.

- **Think about style**

  Another way to use reverb is to create a sense of style throughout the song.

  If you use a master reverb that you add to all the instruments, it's going to dictate the sense of style and space for the song.

    - Use a small room and you'll get a tighter, rawer sound.

    - Use a big hall and you'll get a bigger feel.

  Depending on what you choose, you can take the song in multiple directions.

  Reverb is not about slapping some space on a track and calling it a day. It's a deliberate and thoughtful process where the space you choose can dramatically change the sound of your mix.

  So you have to create spaces around these instruments that make sense.

You can't have big string pads and violins sounding like they're being played in a garage. And you can't throw a cathedral style reverb on a rhythm acoustic guitar. It'll just clutter up the mix.

By spending the time to go through the reverb settings and finding the right types that fit the song, you'll end up with a better mix.

THE DIFFERENCE BETWEEN PRE AND POST-FADER WHEN USING REVERBS

Many DAWs default to a post-fader or post-pan send setting when using effects. For instance, Logic Pro X defaults to post-fader sends and I believe most other DAWs do as well. What that means is that when you manipulate the fader of the source track, the level that goes to the reverb gets affected in the same way. If you lower the volume of the track, the reverb lowers as well, and vice versa.

Most of the time this is great because once you've found the fader balance of the dry track and decide to send the track to a reverb, the relative balance of the reverb changes as you move the dry track up or down.

However, using a pre-fader send can make for some interesting experiments in sound design and can help you make things really spacious and huge.

A BIG DRUM REVERB

Let's take a drum group as an example.

Once you've mixed the drums as you want them with EQ, compression, and maybe some saturation, you'll want to add some depth to them with a reverb. So you create a drum reverb that's just long enough to give the drums some space, without cluttering up the mix with too much reverb.

So far so good.

Then you send the drum track to the reverb track to add space. Not too much to get washed out. Not so little that you can't hear any difference. A Goldilocks amount. It's just right. And because the send is post-fader, whenever you change the fader level of the drum group, the reverb level stays consistent throughout.

However, there's a breakdown part in the middle of the song where you want the drums to sound weirdly distant. You want the drums to sound like all you're hearing is the reverb instead of the blend of the two tracks.

One way would be to mute the drum group without muting the reverb, but that's tricky in some DAWs. And if you lower the fader volume of the drum track, the reverb volume goes with it because it's a post-fader send.

■ PRE-FADER SEND TO THE RESCUE

That's when a pre-fader send comes in. By switching your sends to pre-fader, you can control the amount that goes to the send independent of the level of the original track.

With a pre-fader send, you can lower the volume of the drum group all the way down, and it wouldn't affect the send level going to the reverb. That way, you can make the reverb dominate the track as an effect if you want to make something sound distant and washed out.

I use this trick all the time when I can't get the right dry/wet ratio when I'm doing weird effects. It's great for pushing big lead lines to the back of the mix and creating textures in general.

Keep in mind that you should use this trick sparingly. It's an effect, not a way to glue your tracks together. But it can come in handy when you need to push things way back in the mix. In our drum group example above, you could lower the fader volume all the way down in the breakdown, and all you'd hear is the heavily reverbed drums in the distance.

USING ONE REVERB IN THE MIX

One of the easiest ways to add reverb to your mix is creating a master reverb for everything. Once you've found a good preset that works for the style of song you're mixing, it's usually just a matter of adding everything to it. At that point you just decide what you want dry and in the front and what you want to push behind the other elements.

Select a medium hall, room, or plate and set the decay time for about 1 - 1.5 seconds. You'll quickly notice how the instruments react when you add reverb to the mix. A good way to start is to just have a goal of not making it too noticeable. Just try to use it to add some depth.

Say you want to push the drums and any backup vocals behind everything else. Then you add the most amount of reverb to those elements. If they sound too loud because there's so much reverb on them you can just turn down the fader of the dry track.

Pro Tip: If you like heavy reverb on the track but want

the source instruments to be really quiet, this is where that pre-fader reverb technique I talked about earlier can really come in handy. This will help you keep the source instruments really quiet while maintaining the heavy reverb feel and space of the track.

If you've hypothetically pushed the drums and the backup vocals all the way to the back you really only have two points of the front-to-back depth spectrum. Everything else might still sound pretty up close and personal so you'll want to add some more reverb to the tracks that you want to play second fiddle to the lead instruments, like vocals and guitar solos. By adding some reverb to the guitars (but not as much as you add to the drums), you'll essentially push the guitars a little bit behind the vocal while keeping them in front of the drums.

If the vocal is really dry then add a splash of reverb on there as well, just keep the reverb amount less (or smaller) than the other elements of the mix you're trying to push behind it.

I tend to leave the kick drum and bass guitar dry unless I'm going for a specific effect.

Of course, this is only one simple method of doing it, but it's a good starting point.

And here's another important note: once you add reverb you also need to rebalance your tracks accordingly. Reverb changes the makeup of the mix, so constantly going back to the original fader balance is crucial to keep things together.

■ ADDING SEPARATE REVERB STYLES TO DRUMS

You may want to add a completely separate reverb to the drums that is different than other reverb styles in the mix. And you might even want to put a separate reverb on the snare in order to give it some extra space and thickness in the mix. The drums often need a lot more finesse in the space department to make them sound spacious while not cluttering up the rest of the mix.

Sometimes it's enough to just add a nice "drum room ambience" to your drums and call it a day. A preset like a "drum room" or a "drum plate" can be a great starting point.

However, in some genres you really want to bring that snare into the forefront and make it sound a little different. A good way to get a good, long snare "thwack" without adding too much extra reverb decay is to find a really good short reverb sound that adds more density than "space" to the snare. This is when you're using reverb to add "tone" rather than "space" to an instrument, something you can do with other instruments as well. Find a reverb that's short enough to just give that extra smack to the snare drum without cluttering up the rest of the drum track. This snare reverb will bleed into the rest of the tracks, especially if you didn't gate the snare drum. This doesn't have to be a bad thing so just A/B the sound. Maybe that tight extra room on the snare gives the drums some extra power. So much of mixing can be accidental so experiment whenever you're not on the clock to finish a mix and start taking mental notes of all the weird, unusual tricks you pick up.

If you want to have absolute control over the snare sound it's probably a good idea to take the snare track out of the drum mix group. That way you can mix it completely independent.

■ SOMETIMES THAT ONE REVERB JUST DOESN'T CUT IT.

One of the things to think about when you're making reverb decisions is that your cool reverb might not work in every part of the arrangement.

For instance, if you have a big reverb on a big chorus it might sound out of place if the verse is really calm and chill. Think about it. What if the really huge chorus transitions into an intimate verse and the snare track is still sounding huge with a big reverb decay? That just doesn't work.

A good way to make space and reverb work in a loud-to-soft arrangement is to mult the tracks and process them separately. For instance, once you've got a great snare sound with EQ and compression you might want the same general sound in both the verse and chorus. The only difference is the blend of the reverbs. By having two different tracks that send to different reverbs (a verse reverb and chorus reverb) it'll be much easier to dial in the exact amount of reverb you need for each part of the arrangement.

Another way to combat this is to use automation to send varying amounts of the instruments to the reverb so that drier sections stay dry while the big sections get automated to sound big.

EQ YOUR REVERBS

EQ doesn't only belong on your recorded tracks. EQ is also extremely useful for making your reverb fit better in your mix. Use EQ to avoid unnecessary low-end clutter or reduce the highs to make the reverb blend in better.

TWEAK YOUR REVERB PRESETS

The funny thing about all those presets on your plug-ins is that they have nothing to do with what you're using them on. They're a generic, best-fit sound that might work on something similar to what you have in mind. How's that for ambiguous and vague?

However, presets create a good starting point. Reverbs have enough parameters so that you can easily tweak them to fit your session.

DON'T OVERDO IT

I probably still do this sometimes. This is the classic beginner's mistake. Too much reverb drowns out your mix and makes everything sound "floaty" (for lack of a better term).

Scale back on the reverb. Use it to make your tracks stand out, but don't make the reverb be the thing that stands out. Makes sense? You want to listen to the great sounding vocal. You want the reverb to be an inherent part of the vocal sound. You want the reverb to draw attention to the vocal, not itself.

DON'T ADD TOO MANY SPACES

Too many different spaces in a mix can sound unnatural. If that's what you're going for, disregard what I'm saying. Otherwise, try to stick to a just few spaces. Combining different instruments and tracks into the same reverb works well to glue your elements together.

Using only one universal reverb might work in some cases, but it's a little limited, especially for modern productions. Everything will sound too similar and nothing will stick out. It's a delicate balance of a few different reverbs used in moderation that is the key.

◼ USE THE THREE-VERB TECHNIQUE

My three-verb technique is a simple exercise in creating limitations that you can still work within.

Here's how it breaks down:

- **One Short Effect** – Make this a really short reverb, or maybe even a delay.
- **One Medium Reverb** – Depending on the genre of the song and the BPM, the length and reverb mode may vary, but something like a short 1.2 second plate, chamber or room can work well.
- **One Long Reverb** – This reverb will be used sparsely to push instruments back in the mix so a large hall or chamber can do the job nicely.

Use this reverb technique to create a blend of spaces in a mix without overloading your tracks with too many reverbs.

It's an easy method to place your instruments wherever you want them in the front-to-back field. You'll be surprised at how easy it is to create the necessary space and depth without creating a different reverb and effect bus for every single track in your mix.

You could even complicate the routing a little bit further by sending one reverb to another.

For instance, if you sent the delay to either of the reverbs through a pre-fader send, those delay repeats would sound diffuse and smooth. Or if you sent the long reverb to the delay you'd end up with a large repeated reverb, something that might sound better as an automated effect rather than a sense of space. Experiment with this technique in your next mix to get a feel for how it all works. And if you need multi-tracks to play with, check out the *Step By Step Mixing* resources at www.StepByStepMixing.com/Resources.

■ USING REVERB ON GUITARS

One of the things to think about with guitars in general is that they can totally benefit from a nice reverb, but the more rhythmic they are the more trouble you might get from the reverb cluttering up the actual performance.

You don't necessarily always have that problem, but keep that in mind when working with funky fast-paced guitars that really need all the definition they can get. The more reverb, the less definition you'll get out of the notes because they'll be drowned in space and reflections.

Additionally, if you're working with two different guitar parts, one sustained and one rhythmic, you might not want to add the same type of reverb to them. Using too much reverb on the rhythm guitar can add low-mid muddiness and clutter up the mix, or make it sound too washed out. If you do send more than one guitar to the same reverb, try sending more of the electrics than the acoustics.

Of course, if you feel the guitars are too reverb-y in other parts of the song, you might want to automate the sends to make them fit with the different dynamics. And if you feel like it's nice but too cluttered, sometimes reducing the return of the reverb can make the guitars have the same amount of reverb sound to themselves, but there's less reverb in the actual mix. It can make the elements fit better together. But again, that's all about rebalancing.

The thing about reverb is that there's really no one way to do it. There are guidelines and methods to make things simple, like my one reverb technique outlined above. But ultimately it's about experimenting with the reverb plug-ins you have at your disposal and just finding the right space that works with the style or arrangement of the mix.

ACOUSTIC OVERCOMPENSATION

Adding too much or too little reverb is not always a byproduct of bad mixing skills, but bad room treatment. A bad sounding room can have all sorts of effects on your mixes, like adding too much low-end or too much room reverb. It has nothing to do with you being a terrible mixing engineer.

Sometimes you just have to fix your room. We'll talk about "translation" soon, but make sure you're adding adequate reverb to the real mix, not to the mix that's playing in your specific room/studio. Check it on other speakers and in other rooms. A good way to get a gauge on the "real" amount of reverb in the mix is to check it out on headphones, which we'll be talking about in later chapters.

WHEN DELAY IS BETTER

Maybe you want to let go of the reverb for one day and use a delay instead. Delays are easier to handle, and some are much less confusing than the average reverb.

Sometimes you just need a little depth, without adding reverb, and delay can easily do the job. Delay is a pretty simple processor. Whenever you send an audio signal into a delay it simply delays it for a predetermined amount of time and then blends it with the original source signal. This will usually result in some form of an echo. If the delay time is really short, you'll hear it close to the original signal. If the delay time is long, you'll hear it echo long after the original sound source has died, similar to if you're listening to your voice echo off the cliffs on the other side of the Grand Canyon. Then, depending on the amount of feedback you use, the processor will repeat the delayed signal multiple times.

■ DELAY PARAMETERS

These are the most common delay parameters you'll find on every delay. Some delays might have more analog modeling or "character" styles, but the following parameters are fairly universal across all delay plug-ins.

- **Delay Time** – This is the time you set by which the signal gets delayed. Delay time is either set in milliseconds, so you'd hear the delayed signal a certain millisecond time later, or in sync with the beat of the song. For instance, if you sync the delay to the beat of the song you can make the delay sync in different beat divisions, such as 8th notes or quarter notes, resulting in a more musical and rhythmic delay.

- **Filters** – Delays usually have some basic EQs that lets you filter out the highs or the lows of the delayed signal. This is useful if you don't want to clutter up the low-end (using a HPF) or need to darken the delayed signal (using a LPF).

- **Feedback or Repeats** – You use this control to affect how many times you want the delay to repeat, or feed back around again. If you just want some extra depth, a feedback of zero or only one repeat will add thickness. However, you can get creative with a lot of feedback and infinite delays by cranking up the repeats. A word of warning however, there's a reason it's called "feedback" because if you turn it up too much it will eventually over-load everything and possibly even hurt your speakers if the delays are too loud.

- **Modulation** – Many delays have modulation parameters like Rate and Depth. This will modulate the delay similar to how a chorus plug-in would do it, slightly detune and modulate the delayed signal compared to the original source.

- **Phase** – This button inverts the phase of the delayed sig-nal.

- **Diffusion/Tape Head** – Sometimes delays have analog modeling built-in that's designed to make the delay a lit-tle more analog-sounding. Sometimes this dirties up the signal slightly (similar to saturation), or offsets the de-lays as to not be perfectly aligned on the grid.

- **Mix** – The mix setting allows you to decide how much of the original signal versus the delayed signal you're put-ting through the track. If you're using a delay as a send effect (which is usually recommended), then you'll want this at 100% wet, or only the delayed signal. Because the original sound source is coming from its own track and the delay is on a separate auxiliary effects bus, there's no need to add more of the original source to the delay track.

▩ DELAY DOESN'T HAVE TO BE COMPLICATED

I find that some delays can get a bit too complicated, especially when you just need something simple to work with.

It's best to sync a delay to the tempo of the song. Set it to 8th or 16th notes and blend it underneath just to give a track some depth without it noticeably standing out. For instance, with lead vocals I usually like to thicken them up and give them some depth. When you do that you also need to make sure the feedback is very low.

You only want it to repeat once or so, almost indecipherable from the main vocal. Making the repeat value high will just make the delay feedback repeatedly and it'll sound terrible. Of course, if you're going for a specific effect, or automating the feedback to create a transition of some sort, that's a whole different ball game.

EQ-wise I'd recommend filtering the lows and highs. Usually delays have convenient filters for that. Finally, the LFO (low frequency oscillator) Rate and Depth help slightly detune the delay so it's not a perfect clone of the original track. This will make it sound a bit more like automatic double tracking and it'll add more depth to the vocal track.

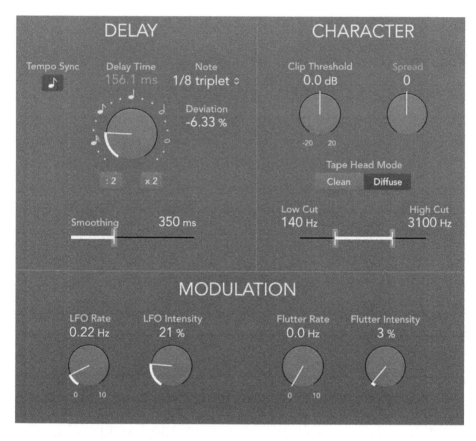

Mono Tape Delay with slight modulation and filtering in the highs and lows

Those are really the only parameters that matter when I use delay. There are a lot of crazy, complicated delays out there that might be used for different purposes and for different genres, but I've found that the simpler the delay, the better I can use it in the mix. For instance, the stock Tape Delay in Logic is very simple and one of my go-to delays, even though I have the Fabfilter Timeless Delay plug-in. Timeless is just too complicated for my purposes so I tend to stay away from it.

So, when would you substitute your reverb for a little bit of delay?

- **For Guitar Solos**

  Sure, guitar solos can sound awesome with a hefty amount of reverb. But they can sound equally cool with a nice delay.

  Use a short to medium stereo delay with one repeat. It'll add width and depth to your signal immediately. The stereo delay will make the solo sound wider, and the single repeat from the delay will add the depth.

  And if you have the original signal in the middle, summing to mono won't ruin the sound.

- **For Rhythm Guitar**

  Using both reverb and delay can quickly ruin a tight rhythm guitar take. If you use too much of either one, you'll end up with a cluttered guitar track.

  However, using a short slap echo or 8th note delay can add interest. Send your guitar track to a delay via a send, then mix the delayed track underneath to add a little space. It doesn't have to clutter the track if you use it sparingly.

- **For Lead Vocals**

  For an in-your-face lead vocal, scrap the reverb entirely and use delay to make the soundstage wider. Delay adds space without making the vocal sound distant, something that happens all too often when you use too much reverb. Depending on the BPM of the song, style, and genre, use either short, medium, or long delays.

  If it's a ballad with long, drawn out words then a long delay creates a big sound without overpowering the actual vocal. A fast rock song benefits from a short, subtle delay and groovy pop songs use medium delays to great effect.

Also, if you have a "verse vocal" and a "chorus vocal" you can easily use send effects to distinguish them better. For instance, if you don't want a big reverb in the verses but you think it would be nice to have a larger space in the chorus, just send the verse vocal to the delay and you send the chorus vocal to the vocal reverb. This will add contrast to the arrangement as well.

- **Adding depth without space to your vocal**

  Use this technique when you want vocal depth without space. It'll make the vocal larger and more present without adding any reverb tail that could clutter up the mix. Send the vocals to a stereo delay with 21 ms on the left and 29 ms on the right. Then use a pitch-shifter to detune or pitch up the vocal about 10 cents. Add the send under the main vocal track until you've achieved the desired ambience needed. An advanced way to do it is with two mono delays panned hard left and hard right with one pitch shifter detuning the vocal 10 cents while the other pitches the vocal up 10 cents.

DIFFUSING DELAYS FOR SMOOTHNESS

Sometimes you want to soften up a delay and make the delay repeats sound smoother. If you want to tame the delays, all you need to do is add a reverb after the delay bus and add a short delay so that every delay repeat will get diffused by the reverb.

- **For Drums and Percussion**

  A similar problem arises from using too much delay on percussion as it does on rhythm guitar. A short delay timed to the BPM of the song gives percussion punchiness without giving it too much room in the mix.

Some genres use delays on the drums to great effect. For instance, a dub reggae snare sound has a short delay and a lot of spring reverb to get that signature sound. Just make sure you keep the feedback low so it doesn't continually echo throughout the measure. You want to keep those rhythms tight. Experiment with delays on the kick and hi-hats as well to take a simple beat and make it more interesting. There's a link in the Step By Step Mixing resources to a great article with audio examples of exactly how to do this: www.StepByStepMixing.com/Resources.

- **For Organs, Synths and Keyboards**

    Medium to long delays with a fair amount of feedback can beef up an organ or pad sound. If you have an organ playing long, sustained chords then a long delay can give that foundation a thicker sound.

    Simple, fast delays work really well on keyboards as well because they seem to like chorus-type effects. Delays can work really well for that instead of using a chorus plug-in.

    Add a stereo delay on a bus and have one side delay in 8th notes and the other delay in 16th notes. Then you can go even further and send that into a modulated tape delay that choruses the stereo delay to create additional "shimmer" to the sound.

DELAYS FOR STEREO WIDENING

A cool way to add some stereo imaging to your sounds without needing the fancy stereo plug-ins is to use panning, delay, then modulation within the delay and some EQ to make the delay sound different than the other source. The best part is it also makes it shimmer more in mono!

Here's what you do:

- **Step 1 – Panning**

  Pan your dry instrument to one side of the stereo spectrum. Then send it to a mono bus that you pan to the other side.

- **Step 2 – Delay**

  Add a short delay of about 30 ms. For added measure use the depth and rate control on your delay to slightly detune and modulate the delay to create further separation. Toggle it on and off to hear how the instrument widens across the stereo spectrum. It's especially great for mono keyboard pads and organs. Oooooh lush!

  But we're not quite done because it's really just a mono panned delay at this point.

- **Step 3 – EQ**

  By using EQ on the delay bus you can make the delay sound different than the original signal.

  In combination with the modulated delay that's panned across the stereo spectrum you get a pseudo-stereo doubling effect that's very effective when you just want some subtle depth in your mixes but don't want to screw around with stereo widening plug-ins.

  EQ is so handy in every situation. Not only do you use it to make all of your instruments fit together in the mix, but you also use it to sculpt your sends to make them work better with your effects.

DELAYS FOR EFFECTS

You can play around endlessly with delay to create weird effects and soundscapes. Play around with the feedback controls to add continuous delay, make each side of the stereo delay different (¼ notes and ½ notes for instance), and then modulate and automate the delay to create weird effects. Combined with a reverb and even some saturation you'll be able to create some crazy textures in your songs.

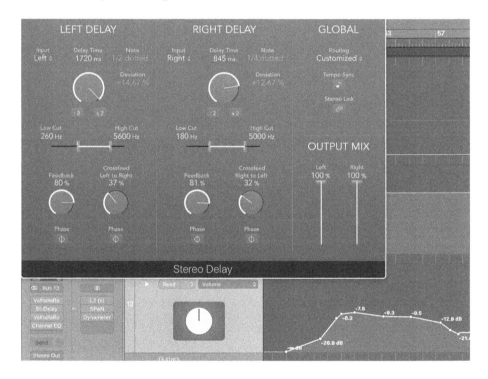

One metal mix I did had a breakdown section where it was just a solo electric guitar and a vocal and it sounded really sparse and empty. So I automated the final distorted power chord of the previous section and sent it through a reverb, a really long, crazy delay (pictured above), and then back into a huge reverb. Then

I automated the volume of the section so it would sound like a noisy texture underneath the instruments, making the whole section sound much cooler than before.

There's a lot of cool stuff you can do with reverb and delay that's not just confined to the standard ways of adding depth and space. Take some time to experiment with your tracks and use these two effects to create excitement and style.

Chapter 6 –
Saturation

Now it's time to talk about Saturation and a few ways to use it on your tracks.

Before we begin I wanted to let you know what saturation is NOT.

Saturation is not distortion. You can use saturation to distort, but by default it's not distortion in the sense that it destroys or dirties up your sounds. Used aggressively it will, but we will not be talking about using it for that effect.

Instead, it can be used to warm and thicken things up, add some grit (but not distortion) to your sound and generally give you some added depth by adding harmonics to your tracks.

There are multiple saturation plug-ins out there. The stock plug-ins in Logic, for instance, include some distortion plug-ins that you could hypothetically use, but they're mostly of the distortion variety, and not the tape saturation and analog tube warmth that I'm talking about.

The plug-ins I use regularly are these three:

- Plug & Mix Analoger
- Fabfilter Saturn
- Waves Kramer Tape

The Plug and Mix Analoger is simple and easy, like most of the Plug & Mix bundle.

The Fabfilter Saturn can get incredibly complex if you want it to be, but it can be super simple as well. We will be sticking to simple and practical because that's my style. It does have multiple types of saturation emulations, everything from clean tape and tube emulation to heavy distortion, so it could be a one-stop saturation plug-in really, but I'd like to compare them a little bit.

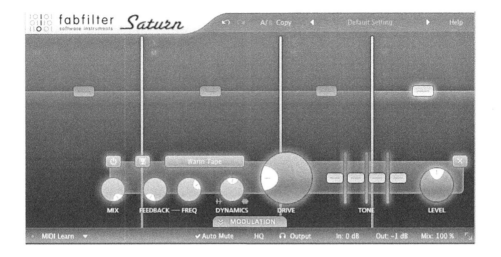

The third is the Kramer Tape from Waves, which is just a tape emulation and it's really fun to screw around with. I use it on many things, everything from the master bus to get some warmth, to individual tracks that need something extra.

But don't despair that these are all paid plug-ins. If you simply Google "free saturation plug-ins" you'll get a whole host of options that you can use in case you don't have any already.

SATURATION SETTINGS

Ok, let's get into these plug-ins a bit. You can review the screenshots above as we go through these plug-ins. You can also download the demos to try out yourself, and if you download some of the free saturation plug-ins available chances are they'll have some of the same parameters.

It's my hope that by covering the parameters of these three very distinct saturation plug-ins that you'll find some similarities between them and what you end up using.

■ THE ANALOGER

The Analoger is the simplest, featuring the lows and highs knobs and a less and more knob. You can select your saturation type from tube and tape or mixed, and then you can choose how much you want to mix it in with the dry signal. The mix knob is great because you can dial in a really saturated sound, but if you only have it set for around a 20% mix you can blend it really nicely with the original sound.

- **The highs and lows knobs** are just how much you want to saturate the highs and how much you want to saturate the lows. Simple and efficient.

■ KRAMER TAPE

The Kramer tape has a few more buttons, but it's purely a "tape emulation".

- **The speeds** are for whether you need it for enhanced low or high frequency response. In short, the low tape speed is good for low frequency instruments, but it could also be used to dull the high frequencies a bit if you needed that. This is because the lower tape speeds result in some high-frequency loss.

- **The input** defines whether you want to hear the signal as it reaches the recording head, or if you want to hear the output as it goes through all the parameters like the tape speed, bias, flux, and all the other settings in the plug-in.

- **Bias** is sort of like tape overload, kind of like adding a bit more gain to the tape signal. In electronics it's a DC off-set so that if you add a certain amount of bias, the top of the waveform will hit the edges before the bottom of the waveform. On some plug-ins, this acts as additional saturation, which can give you good or bad results depending on the track.

- **Flux,** according to the manual (which doesn't really tell you any practical uses for it), is "essentially a gain factor reflecting a higher level passed onto the record head", so in a way it adds a bit more gain or distortion through to the output.

- **Noise** is just that, noise. I'm not a big fan of that because if you use a lot of noise on multiple tracks you'll get an accumulation that'll make your mix real hissy.

- **Wow and flutter** make the signal sound a little more rough, or worn. Basically making the tape sound older.

Then you have the VU meters, and I try to be careful to do correct gain staging so that it doesn't overload unless you really want it to. I tend to take the "Join" button off so I can calibrate the input and output independently if I need to.

The delay section is super convenient for quick spaces. I tend to put this tape plug-in on lead guitars and just use the tape delay to create some additional depth. You can get everything from fast slap echo to nice delays.

■ FABFILTER SATURN

The Fabfilter Saturn is a favorite of mine because it's incredibly versatile. I won't go into all the settings, but here's my quick-start guide for what I usually do.

I obviously put it on a track, then I select the saturation emulation I'm thinking I want. Usually this is some form of tape or tube, depending on the track. Then I find the optimal saturation in the drive knob and by playing with both the drive and mix knob I can usually get close to where I want to be.

The feedback knob isn't usually helpful to me because it basically sends the output signal back into the input and can create some cool sounds, but for the type of stuff I mix I don't tend to find much use for it. You can sort of think of it as continual distortion.

The dynamics knob works kind of like a built-in compressor for your saturation.

The Saturn also has this awesome four-band EQ that can really help shape your sound. For instance, this can work nicely when you want to add saturation to the bass to bring out the lows but also want to make the middle frequencies cut through the mix. Then you simply boost a bit in the middle band to accent that frequency.

One of the more convenient functions on the Saturn is its multi-band capability. It lets you divide the frequency spectrum into different bands, making it a multi-band saturator. This comes in handy when you want a bit more dirt on the lows but would like to keep the highs clean.

APPLYING SATURATION IN YOUR MIX

Warning: One Size Does Not Fit All.

The reason I mention these three different saturation plug-ins is because they all react differently to whatever track I put them on. Even if I put the Kramer Tape on the bass in one mix, it might not necessarily give me the same great results in the next mix.

I highly recommend testing multiple saturation plug-ins to see which one reacts the best to your particular tracks. If you give up and abandon saturation just because it didn't work one time, you're cheating yourself out of a world of possibilities to get a warmer and thicker mix.

- **Drums**

  For the kick drum it's a good idea to try some simple saturation if you need to beef it up in the low end. For instance, if you were to use the simple Analoger, the highs and lows knobs are crucial because too much lows will muddy up the sound, but you still want to crank the highs a bit to accent the top end of the kick. Then it's a matter of working the drive and mix knobs to get the desired sound. Usually plug-ins have a preset that can give you a good starting point to work from.

The same goes for the snare drum if it's lacking in the low-mids and you want it a little more powerful. As a rule, saturation tends to dull the highs a little bit. If you have a really crisp snare that you want to reel in and add power to, it's a good idea to check what saturation can do to help. However, having a multi-band saturator like the Fabfilter helps you add different styles to different parts of the frequency spectrum.

Say you want to add some more meat to the mids without adding more saturation to the high-mid crack of the snare. Then you split the frequency bands down the middle and apply a different saturation style to each band. Alternatively, if you don't have multi-band capability you can duplicate the snare track, add an EQ before the saturation plug-in on the extra track and filter out all the frequencies you don't want to be affected. Then just blend the two tracks together.

Sometimes saturation sounds a little rough in solo, but if you play it in the context of the mix it can really add to the character of the sound.

Because of the nature of some saturation it might not always work on drum overheads. If you want crisp cymbals that cut through the mix then leave the saturation out. However, if they're piercing your ears you can tame them with a simple tape emulator.

- **Bass**

 Like the kick drum, bass guitar can often benefit from the added thickness that saturation gives a track. I tend to gravitate towards my Kramer Tape plug-in for bass if I need some additional thickness in the lows. I start off with a preset, like "bass fingered rock" which is one of their presets, and then tweak the input and output so I can hear the change better when I bypass the plug-in. You want to make sure that there's equal loudness between your source signal when the plug-in is bypassed and when the plug-in is on. If the saturation (or any plug-in for that matter) adds considerable volume to your signal you'll automatically think it sounds better when in fact it just sounds louder.

- **Guitars**

 Adding saturation to guitars can add extra thickness in the low-mids if they're sounding a bit thin. It can also reduce harshness if you have very distorted guitars with too much presence in the high-mids.

 If you want clean-sounding acoustic guitars, I would leave saturation alone. If you're just looking for a little extra warmth, a tape plug-in can help strummy guitars sit better in a mix. Again, using a multi-band saturator can be helpful if you want the acoustic guitar to poke through in the high-end while warmly sitting with the rest of the instruments in the mids.

Multi-band saturator with gentle saturation in the lows, warm tape in the mids and clean tape in the highs

Sometimes, saturation can actually help tame some of the higher frequencies. So instead of using EQ, you can slap on a saturation plug-in instead. It'll help you tame those higher frequencies on harsh sounding electric guitars (wire-y Stratocasters spring to mind) while also giving you some smooth tape or tube saturation to add some nice harmonic content to your sound. It's a richer solution than just an EQ cut and your mix will get more out of it.

There are times when you want to go a little overboard with the saturation on acoustic guitars, especially if you're actively trying to make them sound less like an acoustic guitar. One time I was working on an acoustic rock EP and I mixed it "clean" without any crazy processing or effects. The musician had some feedback and ideas to try and before you knew it, we were experimenting with amp simulators and saturation on the acoustic guitars. We made them sound less like clean acoustic guitars and more like grungy garage-rock rhythm guitars. And you know what? It totally made the sound of the record more exciting and fun to listen to.

- **Keyboards and Synths**

  Depending on how clean you want to keep the keyboard parts in your mix you might approach saturation differently. Electric keyboards can often sound naturally over-driven, especially if they're recorded through an amp. Adding some saturation, bordering on overdrive, can give a keyboard part a really cool role in the mix. Adding a good dose of wow and flutter helps give it that old-school keyboard sound that I tend to like, and the right amount of saturation can often help it cut through the mix in the mids.

  The same goes for software synths if you're doing any sort of electronic music. By adding tube saturation to average sounding software synths you can often beef it up enough to give it a completely new and fat character.

- **Vocals**

  With the vocals you should experiment with as many styles of saturation as you can. Depending on the genre, you might go for anything ranging from subtle tape warmth to all out distortion. Oftentimes a subtle approach helps things sit better in the mix and adds thickness to a thin vocal.

  When deciding on whether to use saturation on vocals, think of whether you want to use the saturation as a subtle warming enhancement to add depth, or as an audible effect to add grit and distortion.

  If you're looking for subtle warmth, a tape emulation or a subtle saturation plug-in on the insert of the vocal track will add a little extra character to the vocal. That will usually be enough.

However, if you want to go overboard and distort your vocals, do it on a parallel bus so you can finetune the balance between the clean and the distorted track. Believe me, it's easy to go overboard and not only add too much distortion, but accidental digital clipping could occur as well if you're pushing your levels too hard.

SATURATION ON SUBMIXES AND BUSSES

Finally, I think saturation works wonders on busses because it tends to glue things together really well. The key for bus-mixing with saturation is to try to be subtle because you're adding a lot more signals to the busses and your saturation plug-in might overload more easily.

■ DON'T GO OVERBOARD ON SATURATION

I've probably repeated myself a little bit in the past few paragraphs because it's all so subjective based on what you're looking for. And sometimes you might not be looking for saturation at all. So don't worry if you feel like it doesn't sound good on a specific track. You don't need it everywhere!

I don't always put saturation on everything, but I tend to at least check if it adds anything cool to the mix. But if it doesn't, I wouldn't hesitate to pull it off again immediately. That's one of the thought processes you should always have when you jump into the mix. Instead of slapping plug-ins on for no reason, there's a lot of critical listening that needs to be done and A/B'ing and level matching so that you know you're making things actually sound better.

SUMMARY

Let me finish this chapter with a quick story about the time I went to Stockholm, Sweden.

Did you know that the Blue Hall of Stockholm's City Hall is really red?!?

This might not be interesting information except for the fact that this hall is where the annual Nobel Prize Banquet is held, one of the most prestigious events in Sweden, and possibly the world. And not only is the Blue Hall actually decorated with red brick, but the bricks are even chiseled down to make them look imperfect!

As our tour guide told us, which is corroborated by Wikipedia:

> *"The hall was originally supposed to have been plastered and painted blue, a color scheme that would have resembled the water of the bay. But Östberg [the architect] changed his mind during the construction of the hall after he saw the red brick. Though Östberg abandoned his blue design in favor of the unfinished red brick, the name "Blue Hall" was already in general use and stuck." -Wikipedia, (brackets mine).*

Identity crises of architects aside, what astounded me more was what the tour guide told us next. When the brick was put in, it looked too new for the feel of the room. So, in order to make everything old and regal, the workers were commissioned to chisel the bricks down to get a weathered old look to them.

And because of how my brain works, this instantly reminded me of how we use saturation to make pristine tracks sound grittier. When we're mixing, not everything should be clean and pristine. If we're going for grittiness, saturation helps us get the feel we're going for. We might have a perfectly recorded vocal sound that's clean and bright, but for some reason that type of sound doesn't fit the genre of the music.

So, we add some tape saturation and dirty up the low-mids and smooth out the high-end transients, making it grittier while fitting into the mix better.

So the next time you need to replace your beautiful blue track with some raunchy reds, slap that saturation plug-in on and see if you can't get the grit you're going for.

If you've been mixing your tracks while following this book you might notice that the only thing left to do is simple automation here and there, and possibly do some tweaks to some settings after referencing the mix on multiple speaker systems and such. Remember that 20% of the effort gives you 80% of the results, so even though you might have all these different plug-ins being advertised every day, you don't always need to jump on the next special. In a pinch you could easily make a quality mix work with just one EQ, one compressor, one reverb, one delay, and one saturation plug-in.

Chapter 7 –
How to Use a Reference Mix

You might have heard the advice that it's always a good idea to check your mix against a reference mix.

But what does that mean? What kind of reference mix? How do you go about doing that?

WHAT IS A REFERENCE MIX?

A reference mix is a commercial quality mix that's usually been mixed and released in the same genre or style that you're mixing in. It's a mix that you know well, that translates well on different systems (especially when you're testing PAs in live sound), or just something you think sounds great.

It's a mix you select based partly on your tastes but also on the quality of the mix. Although "quality of mix" can also be subjective, the translation, separation, and overall characteristics of the mix aren't that subjective.

Using a reference track gives you a different perspective on your mix. It helps you find out what's lacking from your mix and reveals inconsistencies that you otherwise wouldn't have noticed.

HOW TO CHOOSE A REFERENCE MIX

You find your ideal reference mixes by listening to music and making notes of which songs sound great to you. It's as simple as that. Granted, it can be quite daunting to go through every possible song. But as long as you can narrow down a few good songs in the genres you work in, you should at the very least have a few good references to start from.

Honestly, I have problems with references. I'm lazy when it comes to referencing my mixes, and it's something that I've tried to be better about at this point in my career. But I also think choosing references should be a personal journey. It's fun to hunt for great sounding music.

■ RULES FOR USING A PROPER REFERENCE

Make sure that you're using a good reference mix from the start. If you think your mix is bad, it won't do you any good to compare it to another shitty sounding song.

- **No MP3s** – It's not enough to just listen to one of your favorite songs and compare. You need a quality version of your song. No MP3s, AACs, or any other compressed audio format.

  Only lossless WAV/AIFF or better is good enough. Usually, the high-quality master is the last stage of the professional production, so only use quality audio for a proper comparison.

- **Know Your Reference Mix** – Make sure you know the reference song well. You need to be familiar with the song and know how it sounds on your system and others. Pick something you're comfortable and familiar with.

If you lack ideas on what constitutes a great mix, Ian over at Production Advice has a great post on his favorite albums.

The best part is that you can indulge in all your guilty pleasures in the name of "work" or "research."

> *"Sweetie... I have to listen to Purple Rain 15 times a day. I need to get this glassy guitar sound right..."*

> *"Honey... I'm sorry you don't like the Crash Test Dummies but Tom Lord-Alge has a snare sound you can't beat!"*

And so on and so forth…

You can also get a reference mix from the band or musician you are mixing. Ask the band what kind of music they like, and if there's a specific record they want their songs to sound like. That narrows the selection process immediately.

A WORD OF WARNING

Don't confuse a commercial song with a great mix. They don't always go hand in hand. I've heard great records that don't really sound that good. But the songs are cool, so I like the album.

I would never use them as reference mixes though. So to use a reference mix effectively, you have to select a great sounding mix.

Preferably one that's almost unattainable because it'll challenge you to get better. Your mixing skills will improve much faster if your goal is harder to reach. If you're using a subpar mix as a reference, it'll be too easy for you. You might feel good about yourself for a minute, but your skills won't improve.

So the three rules to picking reference mixes that make your mixing skills better are:

1. Pick an actual great sounding record.

2. Pick the best sounding song on that record. (Remember that even though the single might be your favorite song, it is rarely the best sounding song on the record.

3. Make sure the record you pick sounds so good it's almost unattainable to reach it. It pushes you harder and makes you a better mixer.

Once you've found a reference mix you like the sound of that's in the genre you're currently working on, here's what you do next:

IMPORT YOUR REFERENCE MIX

To use a reference mix, simply import the song into your DAW, on a new track. Make sure you don't have any compressors or mastering plug-ins on the master fader because they will interfere with your imported track.

Lastly, make sure that the song is level matched against yours. That is, your reference needs to be at the same level as your mix.

CRITICAL LISTENING

Then you listen and compare, doing the same critical listening exercise as you did when you started mixing your own song.

What do you hear? How is the mix layered? What stands out?

Analyze your reference mix and make critical listening observations.

Think Tall. Wide. And Deep. (Remember this from our visualization exercise earlier?)

- **Tall – Frequencies**

    Compare the frequency spectrum of both mixes and see if anything stands out. Use an analyzer to help you spot problematic areas. The analyzer will help you see where your mix differs when it comes to EQ.

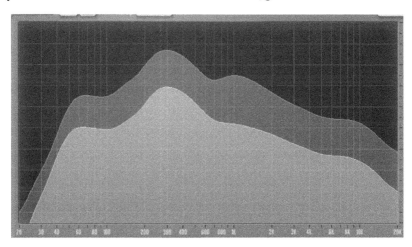

There is definitely something wrong about the lack of low-end and the weird peak in the low-mids around 300 Hz

- **Wide – Panning and Stereo Width**

    Compare the width of both mixes. How wide is each mix? Are the drums panned hard left and hard right, or does the kit sound narrow? Does each instrument have a specific spot in the stereo spectrum, or are they panned to many of the same places? Does the mix have a lot of stereo effects spreading the instruments all over? Are the kick, snare, and bass steady in the center, or is there some creativity used in panning?

The same goes for you. Don't place everything in the middle; instead, spread it out and find a good spot for each instrument to match the reference mix you like.

- **Deep – Effects and Reverb**

  How wet is the overall mix? Are the drums in your face, or are they pushed back with reverb?

  Did the mixing engineer of the reference mix use reverb on the vocal, delay, or a combination of the two? Does the song sound like a band recording in the same room, or is it full of artificial synths with separate reverbs and delays? How is modulation used?

  Push the instruments back, pull them forward, move them closer, or push them away. Think of the mix like a 3D image when you try to match your mix up with the reference.

RECREATE THE MASTER EQ

One of the hardest things to recreate from a reference mix is the overall EQ. For example, say your reference sounds punchy in the low-end and clean and clear in the high end, but your mix sounds muddy and flat.

The proper way to fix your mix is to go back to each instrument and see where the problem lies. Find out where the muddiness is, cut out the boxiness and troubleshoot by using your reference mix as a guide.

Another, simpler way to do this is to use Match EQs. The Match

EQ plug-in in Logic listens to your reference mix and allows you to apply that EQ curve to the master fader of your mix. It's great in a pinch, especially for mastering purposes when your instruments sound great but the master needs some EQ'ing.

However, using a matching EQ is a bit like cheating. I'd prefer that you learned to understand the frequency spectrum so that you instinctively know where to boost and cut to match your mix to the reference you're trying to achieve. That's why I recommend you check out *EQ Strategies – Your Ultimate Guide to EQ*.

It'll help you understand how you can sculpt the frequency spectrum of your mix to sound closer to the commercial reference mix you're using.

MAKE YOUR MIX BETTER BY COMPARISON

As you're making all these observations in the reference, go back and try to recreate them in your own mix.

Take what you like from your reference mix and remix your own mix accordingly. Jump back and forth between your mix and the reference to see if you're making progress. As you go through this process, I wouldn't be surprised if your mix improved considerably in quality as you do small tweaks here and there to get closer to the professional track you're referencing.

At this point you're getting really close to a final mix.

- The balance between tracks is great and you can hear everything clearly and every track is at the correct volume.
- You've EQ'd all the instruments so they sit together in the frequency spectrum.

- You've tamed the dynamics with compression and added punch and tightness to all the necessary instruments in your mix.

- You've added space with reverb and delay so that everything doesn't sound up close, raw and in-your-face (unless you want it to!).

- You've added warmth and relevant grit to your mix with saturation so it feels a little more professional.

- You've compared your mix to a commercial mix through reference mixing. Once you've tweaked your mix closer to the reference mix, you should have made some slight EQ and balance adjustments.

Now it's time to make sure your mix translates accurately to other speakers so that no matter where you play your mix, you can be proud to crank up the speakers for the world to hear.

Get my great printable infographic, the Mix Translation Cheatsheet, to help make your mixes sound great on every speaker, even if you don't have pro gear: www.StepByStepMixing.com/Resources.

Chapter 8 –
Monitoring Your Mix

In order to finish the mix there is something that you should always do to get your mix to translate well for anyone that listens to it. You need to listen to your mix on several different types of loudspeakers and take down notes on what jumps out at you that you need to fix.

Making your mix translate to other speakers is incredibly important.

It doesn't really matter if you have the best equipment in the world. If your mixes sound bad on the bad systems, then your high-end monitors mean nothing.

Here are three things you can do to make sure your mix sounds the best it can before you take it to your car or play it on the home stereo. Do them all in succession or go back and forth between them.

1. **Listen in Mono**

    Flip your mix into mono. Everything still sounding good?

    Is the stereo spread a little weird and your guitar delays and vocal enhancer and spatializers all gone? Then start tweaking your stereo effects until you get as close as possible to the sound you had before.

2. **Lower the Volume**

Turn your monitors all the way down until you can bare-ly hear your mix.

Does the mix still sound the same?

Are the instruments that you want to be dominant still dominant? Can you still hear everything clearly?

If not, it's time to fine-tune your faders to make the mix more balanced. Use EQ to add more presence or pull things back.

3. **Switch Your Speakers**

Chances are you have computer speakers of some sort. If you're using a laptop you can easily switch the output to the built-in laptop speakers instead of the monitors.

Doing #1 and #2 with your computer speakers gives you a completely different listening experience and will show you all the little intricacies of the mix you couldn't hear on your monitors.

Computer speakers by and large sound pretty bad. If you can make your mix sound convincing on both your mon-itors and your laptop, then you're 80% of the way to a well translated mix.

Again, these three things are something you need to do when you're done mixing. Personally, my mix process is as follows:

1. I mix in mono on my Behringer Behritone Mixcube, mostly following the steps I've given you in the previous chapters. The Mixcube is a terrible speaker that has no

low-end or highs. It has only one driver and sounds pretty terrible. However, if I can make my mix sound halfway decent on this speaker I know I'm getting somewhere.

2. I flip my mix over to my Yamaha HS-5s and do another round of tweaking. This is usually a rebalancing effort on EQ but I also flip my mix out of mono at this point. Now I can hear the stereo spectrum pretty well. If I get a "whoa! That mix really opens up even though it wasn't sounding bad before" I know I'm on the right track. I'll spend some time on reverb, delays and other effects.

3. Then I listen to my mix on my Focal CMS50s that are coupled with a subwoofer. Now I can really hear all the little things in the mix, as well as all the low-end that's present. Usually, this requires me to tweak the drums, kick, bass and other low-end instruments.

4. Once I feel my mix is done, I bounce it and upload it to Dropbox. I take the dog for a walk and listen to my mix multiple times on earbuds, making mental notes of what needs to be changed.

5. I either tackle the mix right away, or I sleep on it and come back to it with fresh ears. Throughout this process I tend to check the mix with a high-end pair of headphones every so often to make sure nothing is screwy with the reverb and effects.

6. Once I've done my revisions, I usually get feedback from my studio partner before sending it to the client.

7. If the client has any feedback, I change the mix accordingly and send him the final mix.

Notice how many different pairs of both speakers and individual ears the mix goes through before it's done? You don't want to rely on one single monitor set up in one room to make your final mix decisions. You want to make sure your mix translates well everywhere the mix will be listened to.

WHAT ABOUT MIXING WITH HEADPHONES?

People often ask me, "can I mix and master with just headphones?"

Yes. 100% yes. But don't take this lowly writer's word for it, Andrew Scheps, the mixing engineer behind superstars like Adele, Red Hot Chili Peppers, Metallica, and Beyoncé told the crowd at a NAMM show once that he predominantly mixes completely in the box on headphones. So if he does it without much trouble, I think you can too.

However, you shouldn't just end there. You still need to make sure your mix translates. So make sure you listen to your mixes on as many speakers as possible so you can hear how your mixes translate in the real world outside of your fancy studio headphones. Your mix can't just sound good on your headphones. It has to sound good to the rest of the world too. Regardless of whether you're mixing exclusively on headphones or not, the cardinal rule of translation is that your mix has to sound good everywhere.

Another thing, if you mix exclusively with headphones, you're leaving out one part of the equation.

- **The Room**

   The downside of mixing with headphones is that you still

need to run your mix through every single hi-fi system you can find just to make sure it sounds ok in a real room on real speakers. Unfortunately, with headphones, you're going to do that more often if you don't mix with monitors to begin with.

You need the room around you to make accurate decisions. When you listen to music on speakers you're not just hearing the music coming out of the speakers, but also the music reflecting off the walls. Ironically, the room is also a disadvantage, especially if you don't have an acoustically treated room and it sounds reflective and echo-y. If you're working in a subpar situation without a decent studio room, then headphones might be a better bet overall.

- **Good for Details**

  Headphones are really useful for checking details and doing really close critical listening. Think of them like a zoom in tool for audio. When you need to check out your audio up close and personal you solo it and listen to it in headphones. But when you take the room out of the equation, the sonics will change.

  That said, headphone mixing is quite common, and it does adhere to all the same guidelines as mixing on monitors. It's just a little more time consuming during the "perfection phase" because you still need to make your mix translate to every speaker system.

  If you're stuck mixing with headphones because you'd like to mix your hardcore EDM metal next to your sleeping newborn baby, here are some things to think about when you gotta keep it quiet.

- **Use Quality Headphones**

  Use good headphones from a reputable brand that designs headphones for audio work, like Sennheiser, Shure, or Sony for example. There are more brands out there, so it's up to you to do your research to find the best headphones for you.

  I'm hesitant to recommend specific models because it'll date this sentence faster than you'll find a comment troll on YouTube. As long as you follow the rest of the translation steps in Step By Step Mixing and cross reference your mix on other speakers, not just your awesome (but hyped) headphones, then your mix should be fine in the end.

- **Check Your Mix On Speakers**

  Even if you don't have quality reference monitors in your bedroom, I'm sure you can listen to your mix on some kind of speakers. The car speakers. Your TV's surround system. Your dad's old ("vintage") HI-FI he only listens to AM radio on. Whatever you can find. Take copious notes and make sure to fix anything that sticks out during your mix revision session.

- **Be Careful With Stereo Imaging**

  Your headphones will exaggerate the stereo image because you're putting a speaker on each ear instead of letting each speaker blend together in the room. When you're mixing on monitors you'll hear both speakers at the same time with both ears. With headphones your right ear only hears the right side, and vice versa. That might make for some weird mixing choices that you wouldn't otherwise have done if you had mixed with monitors.

- **Open-Back Headphones Might Be Better**

  Because of the aforementioned stereo problem and lack of crossfeed from one ear to the other, open back headphones can be more helpful because there will be bleed from one ear to the other that doesn't happen with closed back headphones.

- **Review Your Mixes on Other Headphones AND Earbuds**

  To get the absolute best translation possible, not only do you want to check your mixes on as many speaker systems as you can, but you may also want to try other headphones as well as crappy iPhone earbuds. You're mixing for the listener, not the audiophile troll on Gearslutz, so make sure your mix sounds great to the final consumer. They are the most important, regardless of what the haters say.

- **Use Reference Tracks**

  Mixing on headphones doesn't change the importance and usefulness of using reference tracks to make your mix sound closer to a commercial mix. It's an even more interesting exercise because commercial mixes can often sound much different on headphones than you're used to hearing them on speakers. Refer to the reference mixing chapter earlier and follow the same process with your cans on.

- **Use Room Simulators**

  Nowadays there are some plug-ins you can use that can emulate the sound of a room through your headphones. Unfortunately, I don't have any experience with using them because I simply check my mixes on my monitors, speakers, and earbuds like I talk about in the translation. But I thought I'd mention the possibility for the sake of giving you a complete list.

- **Headphones Lie**

  Whenever you mix with headphones you run the risk of pushing things too far back. Because all the tracks are right there around your ears, they automatically sound louder and more present, which in turn makes you add more reverb to push things back. Or you lower the volume of the vocal because it's too present in your mix.

  But the thing is, when you play it back on speakers, you might end up with a wishy-washy mix. Things that were present before might be drowned out by other instruments. The vocal might sound super present on your headphones, but once you play your mix through the monitors it'll be crowded out by all the other instruments. It's easier to start your mix on monitors and then tweak them with headphones. But if headphones are all you have, then make sure you check out your mix on as many speaker systems as possible. You don't want that vocal to be drowned out just because you were lazy.

Chapter 9 –
Wrapping Up Your Mix and
Setting You Up For Success

All right, let's do a quick recap of everything we've talked about so far.

In the first chapter we talked about mixing in general and basic balancing and organization. I recommended color-coding the tracks in the mixer as well as grouping instruments together in busses or groups in order to simplify your mixing.

It's crucial to do critical listening to everything that's going on in the song and even jot down some notes, or make mental notes, of what sticks out that you either want to accent or need to work on when you're first listening to the tracks.

For instance, even after getting a good balance you still might need to work with certain tracks later down the line with EQ and compression if they're being problematic in the static mix.

In the EQ chapter we discussed cleaning up and filtering, cutting unwanted frequencies and boosting flattering ones. Filtering is crucial and you should use both high and low-pass filters to clean up each side of the spectrum. Then we did subtractive EQ to take out annoying resonant frequencies and accented the tracks by flattering the nice frequencies with boosts.

After that we talked about compression and you learned some useful tips to try out in your mixes.

In the reverb and delay chapter I gave you a few different ways to add space to your mixes. I hope you took some of that advice and then experimented on your own.

Reverb and compression, to me, are some of the more subjective aspects of mixing. You can use compression five different ways that all sound different but still good, and the same goes for reverb.

Saturation is something I love using to bring out some character in the instruments, but it doesn't always work and you sometimes have to use a few different plug-ins or tweak the settings in order to really get the results you want.

Then finally, I showed you my methods for getting mixes to translate, which I hope you'll find your own variation of. Also, be careful not to keep tweaking your mixes forever. If your client is happy, let them have it. If everyone in your band is happy, release it!

ACHIEVING SUCCESS THROUGH CONTINUOUS PRACTICE

One of my first guitar teachers taught me a valuable lesson about practicing that I will never forget. His name is KK, and he's one of Iceland's most celebrated singer/songwriters. Even if you've never heard his music, you might've seen him play Riley Blue's Icelandic father in the Netflix show Sense 8.

I adore this man because he shaped the way I think about practice, perseverance, and perfecting your craft.

Here's the story:

It was the second week of class and I was struggling to show him the results of last week's practice. It was a simple Bluegrass fingerpicking tune called "Talandi Dæmi."

However, that second week of class was a failure for the simple reason that I somehow thought that I could learn how to play without actually practicing.

I was a punk teenager and was more concerned with being known for playing guitar (and having a cool teacher) than actually taking the time to learn how to play. It's easy to lift yourself up and brag about stuff like that when you don't have to whip out your guitar and show off your skills.

It's similar to being a musician that "works on their music" all the time but never releases anything.

So when KK asked me to show him what I had learned, he was quick to figure out that I hadn't put in the time to learn the little lick that he assigned.

Then, in no uncertain terms, he flat out called me out on my lack of improvement:

"It doesn't seem like you've been practicing."

Of course, I had a hard time defending his accusation because the only way to prove him wrong would have been to play the homework assignment correctly.

Then he told me something I will never forget.

I'm paraphrasing because this was over 17 years ago, but he said something along the lines of,

"There are other boys on a waiting list hoping to get classes with me that will take the time to practice. If you're not interested in putting in the time, then there are plenty of other students waiting to take your place."

That moment shaped the way I think about practice and perseverance ever since.

If you're unwilling to practice and improve, you will be passed over in life. There are always other people behind you in line that are willing to put in the time. Things will rarely be given to you, but when they are, they usually come with a catch. But earning something because you took the time to practice and improve your craft over time, that's something to be proud of.

To get better, at anything, you must practice. You will suck when you start, but it's the continuous improvement you make every day through your practice that develops you as a creative in any field.

Nobody's ever put this into words better than Ira Glass, so I won't even try to:

"Nobody tells this to people who are beginners, I wish someone told me. All of us who do creative work, we get into it because we have good taste. But there is this gap. For the first couple years you make stuff, it's just not that good. It's trying to be good, it has potential, but it's not. But your taste, the thing that got you into the game, is still killer. And your taste is why your work disappoints you. A lot of people never get past this phase, they quit. Most people I know who do interesting, creative work went through years of this. We know our work doesn't have this special thing that we want it to have. We all go through this. And if you are

just starting out or you are still in this phase, you gotta know its normal and the most important thing you can do is do a lot of work. Put yourself on a deadline so that every week you will finish one story. It is only by going through a volume of work that you will close that gap, and your work will be as good as your ambitions. And I took longer to figure out how to do this than anyone I've ever met. It's gonna take a while. It's normal to take a while. You've just gotta fight your way through."
-Ira Glass

■ PRACTICE IN PUBLIC

As a creative person, I practice in public a lot. I perform live with bands. I put creative energy into producing and mixing music for other groups.

But the most important thing I do is write. I've been practicing writing in public for over a decade, and it's still terrifying to let your thoughts escape into the real world. Keeping them bundled up in your brain is easier because you don't have to worry about whether the writing is any good. If you don't release your work, nobody can tell you that your work sucks.

And boy can it suck sometimes. Absolute garbage writing that nobody should ever see.

But if I don't practice in public, I can't get the feedback that makes me improve. You should think of your mix practice the same way. Practice in public. And by "public," I mean that they are released into the world, ready to be torn down at a moment's notice by an internet troll. Even worse, they might just get ignored completely.

Given the two, I'd rather take the troll. Then, at least you know that your work evoked some emotional reaction, however negative it might be. You hope that if somebody takes the time to hate you so much, maybe there are lovers out there that are happy to read in silence.

Get Feedback from Continuous Improvement

Because I'm an analytical creative, I gauge the success of my writing by how many people read and engage with it. At this point, my public practice has gained tens of thousands of email subscribers that read my daily writing at Audio Issues. That's the sort of statistic that keeps me going, even when the creative part of my brain tells me my work is meaningless.

However, I never would've created such a large tribe if I had listened to my inner imposter syndrome and procrastinator. It was the dedication to practicing in public and humbly asking for feedback from you readers that helped me achieve that.

So if you want to get better at mixing, there is no way around the continuous practice you must do to get better. You can obviously buy a book (like this one!) to help you understand the topics better. You can enroll in a course to help you improve, or hire a teacher to show you the ropes.

But ultimately, that dopamine rush of accomplishment you get by investing money in yourself will fade. You can't hire a great teacher and expect to get better. You can only get better with continuous practice over time.

There is no shortcut.

That's why I think it's so important to include some multi-tracks inside my Step By Step Mixing resource page. I don't want you to just become a know-it-all on using EQ, compression, reverb, delay, and saturation. That's my job. I want to give you the materials you need to improve through practice.

And if you can't put in the time to practice, don't worry, there are plenty of people waiting in line behind you to take your place.

I hope you'll use the tips I've shared with you throughout this book to make an exciting mix.

Don't Forget Your Free Resources

If you want even more material on mixing, don't forget to grab all the free resources I've put together inside the *Step By Step Mixing* resource section: www.StepByStepMixing.com/Resources.

Here's what you get inside the *Step By Step Mixing* resource page:

- *Step By Step Mixing* Quickstart video
- Mix Translation Cheatsheet
- Links to over 100 multi-tracks to practice your mixes
- Step By Step Mastering Guide
- In-Depth Frequency Chart for the Entire EQ Spectrum

I'm looking forward to helping you improve your mixes even further!

Thank You

Before you leave and try out all these tips I just wanted to say "thank you!" It means a lot to me that you took the time to read this far. There are so many books on mixing and audio so I'm honored you picked mine.

If you enjoyed this book and loved what you read, please leave a review on Amazon.

That's honestly one of the best ways to let others know how valuable this information is. I really appreciate it!

About the Author

Hi, my name is Björgvin Benediktsson. I help musicians and producers make a greater impact with their music by teaching them how to produce and engineer themselves.

I've worked in live sound doing huge concerts back home in my native Iceland (that's where the name is from), recorded bands in Madrid, Spain (where I went to audio engineering school), and now I produce, record, and mix local bands in Tucson, Arizona

(I get around a lot...).

Through these diverse experiences I've grown to learn one thing:

I love teaching you how to make an impact with your music and audio production.

I've written over 1,000 articles on audio and taught thousands of up and coming home studio producers such as yourself how to make an impact with their music through www.Audio-Issues. com since 2011.

If I can help somebody improve the sound of their music and that helps them get extra fans and exposure I think I've done a great job.

I believe in sharing my knowledge with everyone, and through Audio Issues I've been able to do that. I might not have won any Grammys, but I enjoy being the person who inspires others to make great music while aspiring to learn as much as there is to know (and pay it forward!).

That's what I'm all about. I've learned a lot from working in the industry for over 10 years and I hope my knowledge can become your shortcut to taking your mixes to the next level.

If you're looking for more from me, check out some of my other books and courses here: www.audio-issues.com/products.

Lightning Source UK Ltd.
Milton Keynes UK
UKHW021142150721
387213UK00009B/1870

9 781733 688819